Historical Replica Constructions In Wood And Metal

VIKINGS: VOLUME 1
2ND VERSION

I0169391

By Stephen (Sven) Wyley, Wayne Robinson and Shannon Joyce.

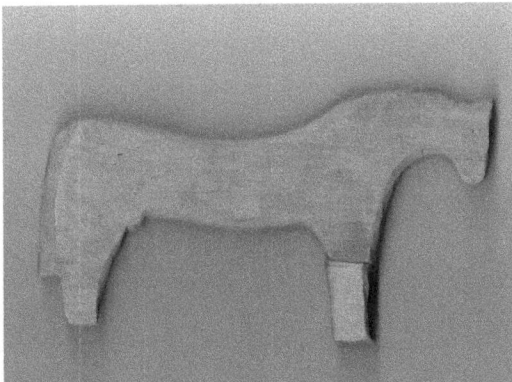

Front Cover: My son's Trondheim toy horse, photographed by my other son Morgan Wyley. Note: the toy horse broke its leg at the same place as the original.

Wyley, Robinson and Joyce. C/- Intertype
Unit 45, 125 Highbury Road
BURWOOD VIC 3125
www.intertype.com.au

Ordering Information:
Quantity sales. Special discounts are available on quantity purchases by corporations, associations, and others. For details, contact the "Special Sales Department" at the address above.

Historical Replica Constructions In Wood And Metal/ Wyley, Robinson and Joyce.. —2nd Version. Edited by Sil Dejong.

ISBN 978-0-6452042-1-6

Contents

Acknowledgements

The following people have contributed over a long period to the development of the knowledge that has resulted in this book; Jenny and Gary Baker, Gary Bannister, Dr Peter Beatson, Dr Timothy Dawson, Stephen Lowe, Hugh McDonald, Christopher (Quarf) Morgan, Steven Nicol, Wayne Robinson, Keith (Chips) Whitehead. Frank Hubert Wyley (my dad), and Andrew Young.

Thanks to following for their assistance:
Anne C. Søensen (Head of Maritime Archaeology, Research and Exhibitions, The Viking Ship Museum, Roskilde, Denmark);
Göran Ekberg (Curator, Archaeology Unit, Swedish National Maritime and Transport Museums, Stockholm, Sweden);
Gunilla Gardelin (Cultural Environment Developer / Curator - Cultural Historical Association of Southern Sweden);
Gunnar Andersson (Senior Curator - Statens historiska museer, Sweden) and
Jenny Kalseth (Norwegian University of Science and Technology).

Thanks for all the feed back and special thanks goes to the following for assistance with the 2nd version:
Sil Dejong, Wayne Robinson, Joshua Button, Derek Rea, Brodie Henry.

I am forever thankful and appreciate all the time, effort and support from the following, which are too numerous to name:
To my friends and family for their love and support;
To the re-enactors that went before me, did the research, wrote the articles, and taught me how to make stuff;
To the archaeologists and researchers that found and researched the extant stuff;
To those that published the papers in places and books which were able to be found and used;
To the many recipients of goods from Sven the Merchant for their support and encouragement;
To all those that participated in the Sven the Merchant workshops, learnt how to make stuff, and then passed on those skills to other people.

Stephen "Sven" Wyley.

"I believe in the sharing of knowledge, not the hoarding of it"

— S.WYLEY

Introduction

The Vikings originated in Scandinavia in countries now known as Norway, Denmark and Sweden. The Viking period began with the sack of Lindisfarne 793 CE and ends with the Battle of Hastings 1066 CE. Vikings were known for being farmers, raiders, traders, slavers, mercenaries and explorers, using their advanced sailing and navigational skills, longships for exploring river systems and across the seas as part of military, mercantile and territorial expansion and settlement, and some even served in the Byzantine Emperor's body guard known as the Varangian Guard.

The geographical extent of the Viking travels included to the East – The Baltic, Kiev, Novgorod, Constantinople, along the river routes; and to the West – England, Ireland, Scotland, and the Orkney and Faroe islands, Iceland, Greenland and Newfoundland (Canada).

Our knowledge of the Vikings is fragmented because of the lack of recorded history. What we have got are rune stones, the writings of their enemies, the writings in the 12th to the 14th century Sagas, linguistic and etymological studies, and the study of the artefacts from the Vikings themselves. Archaeological finds are being revisited and there are still more discoveries being made, adding to our knowledge.

Viking artefacts provide an excellent opportunity to learn how their furniture, equipment and accoutrements were made, used, and functioned, as well as a starting point for learning about other cultures and periods in history.

This volume contains a collection of knowledge, amassed over 30+ years of trial and error, in the pursuit of a better historical presentation. It sets out the why's and wherefores, and highlights the traps and pitfalls for the unwary, so that you the reader can start or continue on the journey to a more accurate presentation for the purpose of living history, re-enactment, a display in a museum or just an interest of history.

This volume covers eight basic Viking projects:

- The Coppergate wooden mallet;
- The Lund stool;
- The Oseberg chest (no. 178);
- The Oseberg bed (no. 2);
- The Sala Hytta table;
- The Trondheim toy horse;
- The Gokstad candle holders;
- The Birka bag hanger.

This book assumes a basic level of experience in woodworking. As you complete the projects your skill will increase, however, safety is paramount. You should not use any tool or equipment without wearing the appropriate personal protective equipment. It is recommended that you get training from a training institution or a more experienced worker of wood and metal. The subject of metal work will be touched upon in more detail in regards to the manufacture, purchase and fitting of metalwork such as chest hinges, hasps and handles.

A definition of "Living History" is the pursuit by the current generation to understand those of the past by attempting to portray and carry out activities of the past without experiencing the adverse outcomes. Those activities may involve the wearing of clothing (in some cases wearing armour as well) and the use of tools, equipment and weapons, which helps the participants to understand the past.

Where as "Experimental Archeology" is hands on way of testing a hypothesis regarding the archeological record to fill the gaps in our knowledge by range of activities like iron smelting, cloth dyeing, pottery, evaluation of armours through making and testing against arrows.

Examples of historical sources range from:
- Extant remains (e.g. Oseberg and Gokstad ship burials);
- Stray finds (e.g. Hedeby chest);
- Illuminated manuscripts (e.g. Cantigas de Santa Mari´a);
- Wills, household accounts and other documents of the period;
- Written eye witness accounts or chronicles of the period;
- Tapestries (e.g. Bayeux Tapestry).

Some of the finds are not complete such as the chests from the Birka cemetery (Grave number 639) [1], where only the metal fittings remain. From them you can work out the size and shape of the chest based on their position in the excavation. Often interpretation of the archaeological reports, drawings and photos is required to fully appreciate the size, shape, and construction methods used which will enable you to construct replicas.

This volume will include the choice of materials and how to store them. The use of various joints and tools used will be covered separately when they appear in each project. So good luck, enjoy the journey, look back at what you have achieved and *"Strive to be happy"*.

---

[1] Arwidsson, Greta (ed.) (1989) *Birka II:3 Systematische Analysen der Graberfunde.* [Systematic Analysis of the Graves Findings].

The Basics

Safety

Personal safety, and the safety of those around you, is of paramount importance. The process of historical replica construction must not put you or others at risk of an incident causing harm or even a fatality.

Some basic rules are:
- Read and follow the equipment makers instructions;
- Use the tools the way in which they were intended (i.e. a chisel is not a screwdriver);
- Wear the appropriate personal protective equipment (PPE) (eye, ear and respiratory protection, safety boots, gloves as appropriate and a leather apron for the torso, no bare skin on torso and limbs);
- Keep tools and equipment clean, in good condition and sharp (if appropriate);
- Provide adequate ventilation in the work space;
- Keep the work area clean, tidy and free of slip, trip and fall hazards;
- Ensure that equipment is firmly affixed to the floor or a bench (i.e. grinders and pedestal drills);
- If standing in one place for long periods, use rubber fatigue mats;
- If you are using power tools ensure that, the power cords are not damaged, the guards are in place and are operational;

- Don't store toxic chemicals in used drink containers;
- Provide adequate lighting;
- Provide and maintain fire fighting equipment;
- Have a Residual Current Device (RCD) on the electrical supply for the electrical equipment being used.

A hazard is a thing that can cause an incident causing harm or damage. The risk is the subjective assessment of level of harm or damage a hazard may cause. A control is a mitigating thing or practice to reduce or remove the effect of the hazard.

Table 1. Hazard identification and control.	
Hazard	Control
Chemicals	Read and follow the Safety Data Sheet (SDS), provide ventilation, PPE (gloves, safety glasses, respirator). Wear long pants, long sleeve shirts. Store chemicals as per SDS.
Drugs, medications, alcohol	Don't work under the influence of drugs, medications or alcohol.
Dust	Mechanical extraction, ventilation, PPE (respirator).
Electrical	Maintain electrical leads and cords. Residual current device (RCD) on electrical switchboard
Entanglement	Tie back long hear. Remove jewellery (or wear gloves. Note: some people believe that gloves themselves are an entanglement hazard. Do not wear loose clothing.
Equipment	Good maintenance, Guards in place. Safety switches operational.
Ergonomics	Set up of workshop (heights of benches and equipment, storage of most often used tools in reach zone, avoid awkward positions).
Fatigue	Take regular breaks (10 minutes every hour). Install rubber fatigue mats on floor.

Fire	Reduce combustible material near ignitions sources (sparks from cutting and grinding metal, or welding).
Flying debris (from cutting, grinding etc.) and sharp objects.	PPE (Gloves, safety glasses or face shield). Long pants, long sleeve shirts.
Heat	Environmental (work in early morning or evening, take regular breaks (10 minutes every hour), drink water). Situations (i.e. forge) (heat shields, PPE (Gloves, safety glasses or face shield). Long pants, long sleeve shirts.
Manual handling	Set up of workshop, store heavy items between shoulder or knee height. Use correct lifting techniques, 2 person lift, or mechanical lifting equipment.
Noise	Distance from source, noise dampening materials, baffles or barriers between source and ears. PPE (plugs (correctly installed) or muffs)
Slips, trips and falls	Remove slips, trips and falls hazards, clean up workshop after each job, good housekeeping.

"Poor housekeeping is a sign of other safety hazards present in the work space"- S.Wyley.

Historial Sources and How To Use Them

When you study and attempt to make replicas based on extant items or even item based on illuminated manuscripts you will encounter the *'evidence barrier'*. This barrier separates us from the past and makes it hard for us to understand how the medieval world looked, let alone how it felt and smelled. The *'evidence barrier'* consists of distance, time, and culture. You need to understand those barriers to replicate it.

Here is a table explaining the relative merit of the different sources of information. The historical accuracy and value decreases as you move from the primary source of the information down to your fellow re-enactors, and the least reliable source of stage and screen.

Table 2: Sources of information versus historical accuracy		
Primary sources – the extant items from the period.		
Secondary sources – carving, sculptures, paintings, drawings, illuminated manuscripts showing the item from the period.	↓ ↓ ↓	The historical accuracy of the information decreases as the number increases.
Tertiary sources – Period writers and commentators writing in the period about the item.	↓ ↓ ↓	
Modern writers (Archaeologist, researchers, etc.) writing about the item.	↓ ↓	
Other re-enactors, the internet, Hollywood.		

It is always the best option to base your work on a primary source(s), and the more examples of an item the better. If there is more than one source which shows the item was used consistently in a certain period, historical context, strata of society, or regional context, this will add more weight to your research.

Issues when using secondary sources include:

- You only get a limited view of the item (i.e. only the front of the item is shown so things like the hinges on the back of a chest are not shown);
- The thickness of the timbers used are not shown;
- Other details (details of lock) are missing or are not shown;
- In some cases the artists of period would use images from their past to illustrate a point;
- In regard to proportionality in art, in medieval art there was a propensity for artists to show the person of interest larger than those around them and not in proportion to other items in the scene depicted.

The more example of a particular item, the more likely it was to be in common use and thus more applicable for a living history reconstruction: 1 of the thing = a good start; two of the same kind of thing = getting better; three or more of the same kind of thing is best!

It should be noted that wood shrinks, warps and degrades over time with exposure to environmental conditions, and metal rusts and leaves encrustations. In some soils nothing but the metal fittings remain of a once grand chest. From the placement of these fittings we can determine the shape and size of the chest. With a certain amount of imagination and referral to other similar extant items, we can render a probable replica of the item. In some cases interpretation will be required and this is a subjective undertaking. A person can only base such interpretations on their own gathered information and the extant items at the time of the research.

Where an item was found may indicate who used it and why. A rich person may have ornate jewellery, a full domestic kit of cooking gear, weapons and shield, carts, sleds, beds and other furniture (such as the Oseberg Ship Burial). The type, finish and number of grave good can indicate not just the wealth but also their occupation or standing in society.

Bad practices.

For historical reproduction and re-enactment purposes the modification or embellishment of an item is an error and should not be condoned, some examples are rope handles or chip carving on chests, as there is no current extant evidence to support these techniques being applied to Viking era furniture. While there is carving on the Hedeby chest, it consists of straight lines around the outer edges of the front and sides, and there is carving on the Oseberg cart, sled etc., but that does not mean you can create a replica Hedeby chest and carve a *"gripping beast"* on the lid, it's just plain wrong.

"Negative evidence is a slippery slope which must be avoided at all costs" - S.Wyley.

Furniture (Form and Functions Of Furniture)

Based on extant remains furniture has forms, functions and artistic design and has been around for more than 8,000 years, some examples from the past:

- 6000 BC – the Seated Woman of Çatalhöyük (Museum of Anatolian Civilisation;
- 3180 BC – 2500 BC – the cupboards and beds from Skara Brae of Scotland;
- 3100 BC – 323 AD – various Egyptian stools, chairs, chest, tables and beds.

Humans developed furniture over time for various function in the dwelling place which included:

- Somewhere to sit = Chair;
- Somewhere to sleep = Bed;
- Some where to eat, play or work = Table;
- And somewhere to store valuables = Chest.

Various other miscellaneous items were developed for art, tools, games and toys.

The size and shape of seating varied a lot. Look at the varying height for the lower leg (i.e. from the ground to the back of the leg, which may vary from the size of the person it was made for and changes in height from Medieval people to Modern times (due to better nutrition and health care)[2]:

- The Oseberg chair (c. 800 – 850 A.D.) (Overall height 67cm, seat height = 33.5cm);
- The bench ends , Hemsedal Church (c. 1200) (Overall height 130cm, seat height = 32.5cm);
- Bede's Chair. (14th c.) (Overall height 150cm, seat height = 45cm);
- Glastonbury chair (1500's) (Overall height 83.8 cm, seat height = 44.4 cm).

---

[2] See Werner, A., (1999). London Bodies The Changing Shape of Londoners from Pre-historic Times to the Present Day, Museum of London.

Time

Time, there is never enough of it.

Most modern people have no real concept of the total time it takes to 'grow a tree' for timber, to fell it, to hew to shape, cut it to size, to join the pieces up to make a chest, or to mine the iron ore, to smelt the ore, let alone to forge it into a knife. As a modern person, if we need steel or timber there is a store for it. The technology and skill of those that preceded us took a certain amount time and effort, and to understand the past we need to experience some of that first hand. So I challenge you, the reader, to cut down your own lumber and square it off into timber to make your own stool, or make your own nails using a forge, tongs, a hammer and a nail header. By participating in activities of harvesting, modification and manufacture of materials you appreciated and understand how somethings have changed, whereas, somethings have not changed that much at all.

"No piece of work is worth making unless it satisfies some purpose".
- Edgar H. Heelas, 'Craftwork in wood', 1944.

Materials

Waste not, Want not

Wood is a resource that is renewable but trees take a long time to grow, so it is important not to waste timber and this can be done by marking out and cutting the timber in an order and manner which produces the least waste. Avoid timber with flaws which can create weak points such as; uneven grain, end split, bowing, warping, shakes, twist or wind, ingrown bark, knots. Set up a cutting list, see Table 4, which sets out which piece should be cut first, to aid in the construction and save on wastage of timber.

Table 4. Showing cutting list for an original Oseberg 178 chest.	
Bottom	64.5cm
End * 2	31cm
Front	66.5cm
Back	66.5cm
Lid	62cm

Nails

Forging nails takes time, and effort, practice, materials and equipment and so if you can do it yourself a black smith will charge you a lot just for a nail, let alone enough for a chest. So the use of 'cut nails' is recommended (produced by a machine that cuts the nails from sheet steel the forges the head) which look like forged nail but they a far cheaper than forged nails.

Predrilling of nail holes is recommended to reduce the chance of wood splitting, using a drill bit a millimetre smaller than the widest part of the nail shank or gimlet (twist auger).3

Photo 2. Some nails that I forged myself.

Glue

Glues have been used for thousands of years to hold things together, including:

- Pitch and resin on Neanderthal tools back to 80,000 years before present;

---

[3] Examples are known across the Viking world, one was found at a dig in Castle Street, Inverness and is now on display in the Inverness Museum and Art Gallery (no find number available, labelled as *twist auger*). https://www.highlifehigh-land.com/inverness-museum-and-art-gallery/

- Animal skin glue over 8000 years old from Nahal Hemar Cave complex in Israel;
- Pitch on the Ötzi man's axe (between 3400 and 3100 BCE);
- Animal based glue was found on a casket in Tutankhamen's tomb (c. 1342 – c. 1325 BCE);
- The Greeks used egg whites, blood, bone, milk, cheese, vegetables and grains for glues;
- The Romans used a range of glues, bitumen birch tar and tallow mix, hide and fish glue, and tar and beeswax;
- And the 9th C Mappae Clavicula mentions mixtures for fish and ox hoof/hide or cheese glues for bonding stones, tree saps and parchment glue.

It is recommended that you use glue that will set, not show and hold the joint good and proper. If you want to spend the time and effort on making animal hide glue and the neighbours or your other half doesn't complain about the smell, make it happen. It sounds like it would be a great experimental archaeology project.

Joints

The earliest joints in furniture can be seen in extant Egyptian furniture: butt joints, mortice and tenon joints, half lap, simple mitre, shoulder mitre, double shoulder mitre, mitre housing and the dovetailed mitre housing, scarf joint with butterfly and housing joints, by comparison most medieval joinery is quite plain.

Joints used in Viking furniture include butt joins (two surfaces abutting each other) and mortice and tenon joints (where a tenon protrudes into the hole (mortice). See Drawing 1 of the various joints used on the Mästermyr chest.

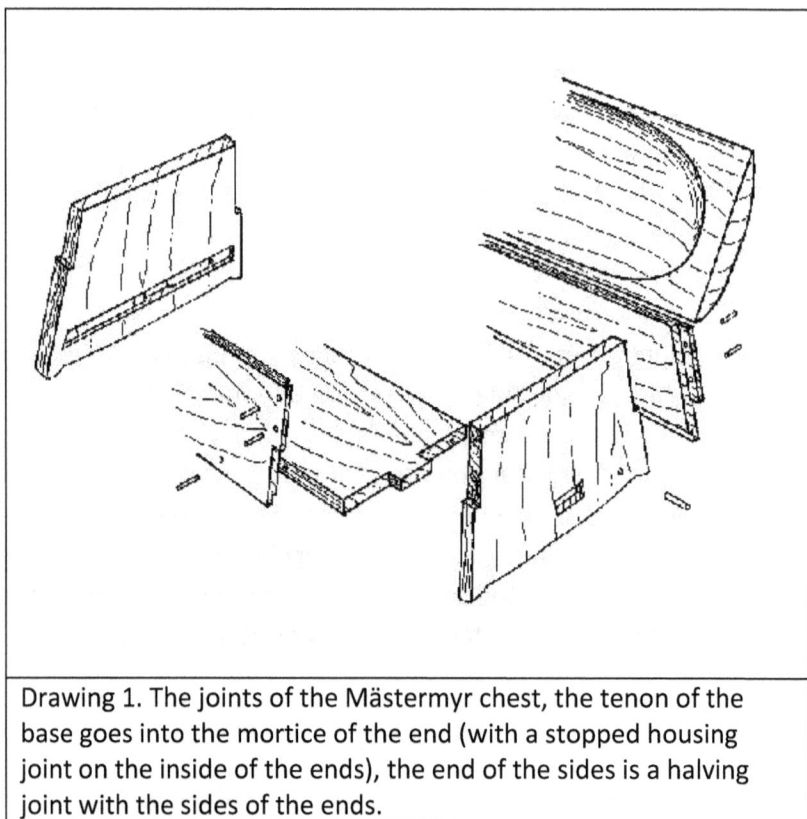

Drawing 1. The joints of the Mästermyr chest, the tenon of the base goes into the mortice of the end (with a stopped housing joint on the inside of the ends), the end of the sides is a halving joint with the sides of the ends.

Finishing off woodwork.

There is no evidence, currently available that shows or proves that Vikings used material like sandpaper on their woodwork. However, Pliny (Roman writer) mentions the use of rough fish skin which was used to smooth and polish wood and ivory.[4] Cato (another Roman writer) says that oils (Cedar, Rose, Juniper and Olive) were used to polish wood.[5]

Theophilus (an artist writing the early 12th century) recommends linseed oil for painting doors and as a varnish. [6] I recommend the use of Linseed oil for the preservation of wooden furniture and other objects. At least once a year, normally in the Spring, you should give them a coating of preservative and allowed to dry. A recipe for Linseed oil is (1 part boiled linseed oil to 3 parts turpentine (you can use mineral turpentine or gum turpentine which is more expensive than mineral turpentine) or beeswax (dissolve equal parts melted beeswax in turpentine, mix well, use for non-food safe items). Use the linseed and turpentine mix in a well ventilated area brush on the preservative and allow to dry.

> **Safety Tip**: Cloth rags which have been used with linseed oil should be soaked in water afterwards because they can spontaneously combust and start fires.

"Good Craftsmanship is the happy combination of good design, sound construction and skillful tool manipulation..." - Edgar H. Heelas, 'Craftwork in wood', 1944.

---

[4] Pliny – *Natural Histories*, 9.14.40: 13.34.108.
[5] Cato – *On Farming*, 98.
[6] Hawthorn & Smith, Trans. (1979), *Theophilus, On Divers Arts, The Foremost Medieval Treatise on painting, glassmaking and metalwork*, New York.

Timber.

It appears that oak was the primary wood for construction and furniture throughout Europe during the medieval period. This may be due to the survivability of oak over other timbers such as ash and willow (See table 3 for a list of other woods found in archaeological digs or extant items). However, in this day and age (an especially for those of us in Australia) oak is expensive and hard to get, especially in wide planks, whereas, pine (Pinus radiata) is cheap and readily available in wide and long planks.

Table 3. Other woods which were used in Europe and they include the following.		
Alder	Elm	Poplar
Ash	Fir	Pomoideae
Aspen	Hazel	Plum
Beech	Juniper	Spruce
Birch	Larch	Willow
Boxwood	Lime	Yew
Chestnut	Maple	
Elder	Pine	

When choosing the timber for the projects select pieces that are; dried, straight, and crack and knot free. This will save a lot of waste, work and swearing later. When you get the timber home store it flat and in a dry place (I use the rafters of my garage and workshop). If you have the time, skill and tools you may harvest and prepare your own timber, for those that don't there are a range of suppliers. Using recycled timber reduces your carbon footprint and there is less chance of the timber warping because older timber has had more time to dry. Specialty timber suppliers exist and you can use the internet search engine of your choice to find the one nearest to you, they have machines to dress the timber down to the desired size.

The use of period tools produces tool marks which cannot be replicated by modern tools, to produce adze marks you need to use an adze.

"The more period tools are used, the more authentic your mistakes become", – G. & W. Robinson.

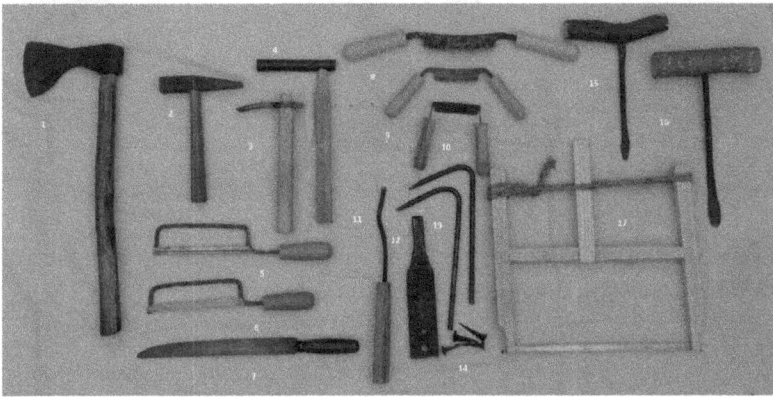

Photo 3. Some period tools, most of the tools pictured could be found in the Mästermyr tool box. 1. hand axe. 2. cross peen hammer. 3. adze. 4. dog hammer. 5. hack saw (fine blade). 6. hack saw (course blade).7. hand saw. 8, 9 & 10. draw knives. 11. chisel. 12. nail header. 13 hold downs, 14. forged nails. 15 & 16 spoon bit drills. 17. frame saw.

Photo 4. The Hook knife produces thin curved lines on a replica of Nydam quiver by Wayne Robinson.

Metal.

Iron bars have been found at a range of Scandinavian sites and most of the steel used in the medieval period had a low carbon content similar to wrought iron. For most of the work required to make the metal fittings for these projects mild steel is the fall-back position as it is readily available and relatively cheap. You should not need training in blacksmithing because the mild steel can be worked cold. Instructions for making various items will be included in later sections. You could also pay someone to make the fittings for you but it could be expensive.Blacksmithing is a worthwhile skill to have under your belt for historical re-enactment or a *"Zombie Apocalypse"*.

Iron was expensive, let alone a sword because of the labour intensive process. If we convert a Saxon silver penny to 20 UK pound, then convert that to Australian dollars, the cost of a sword could vary from $4261[7] to $106,000 (AUD)[8] in 2021.

The process of making iron included finding the ore (i.e. bog iron); smelting the ore; blacksmithing, forging or casting the fittings (bronze casting); making the scabbard (including the wood and leatherwork).

Mild steel comes in a range of shapes (i.e. flat or round bar) and sizes which will only need some modification to be used for fittings. Remember to store your steel in a dry place to avoid rust.

It is recommended that iron and steel of the metal fittings is surface treated (by bluing or blackening) which protects the steel fittings from rust and wear and tear. Blackening involves the application of an oil to the surface and applying heat to carbonise the oil, any oil will do.

The Blackening process (see Photos 5&6) as following:

- Set up in a well ventilated area;
- Prepare flame resistant materials (such as fire bricks) under the work;
- Ensure the piece is clean of dust, oil and other matter;
- Heat up the piece (using a forge, gas torch (LPG, Butane or Oxyacetylene) until it takes the blue/green or brown shade you are after, remove the flame;
- Dunk the piece of metal in oil or brush the item with oil (this will produce smoke so carry out in a well ventilated area);
- Put aside and allow to cool, once cool wipe down with a clean cloth;
- If surface of the item is blotchy you can wash with alcohol to degrease and repeat the process.

---

[7] Lex Ribuaria, a Frankish Legal text from the 7th century.
[8] The Will of Abba the Reeve, Kent, from about 839 A.D.

Photo 5. Blackening chest fittings with oil and flame from LPG torch.

Photo 6. Oil carbonizing, after getting the metal hot enough to turn black but not too hot to burn the oil way.

References to surface treating metal can be found in Theophilus in regards to armour. [9]

---

[9] Hawthorn & Smith, Trans. (1979), Theophilus, On Divers Arts, The Foremost Medieval Treatise on painting, glassmaking and metalwork, New York

Plans and Patterns

It is important to get the best information, I have come across books and museums that have used or provided incorrect dimensions and details. So I recommend you verify it from another reliable source.

1. Obtain plans or dimensions from a good source.
2. Draw up your own plans based on a good source. I recommend a 1:10 (mm: centimetres)(computer aided drawing i.e. cadcam, comes in handy);
3. Draw up in pencil, which makes correcting easier with an eraser;
4. Make a mock up in cardboard;
5. Scan a copy of the drawing and plans so you have an electronic copy, and ensure to make secure backups on a regular basis;
6. Make patterns in cardboard or in thin plywood, write the name of the pattern and the date, store in a safe dry place.

Using plans and patterns make reconstruction of items easier. If you want to make a copy of an item I recommend that you keep a copy of the plans and patterns in a clean dry folder, in plastic sheaths. Also make a scan of the plans and patterns and take photos, store backup copies on a hard drive separate to your computer/laptop etc. Security of data today is as crucial it was in the past to store valuables in a locked chest.

Modification of the plans can be made to fit the end user's requirements (how long is that sword?) and the size of the available timber. If the timber planks you have are not wide enough you can always glue boards together to come up with the right sized timber.

From Manuscripts

Manuscripts written at the time are secondary sources and have to be used with caution; the artist did not always paint the people, or the furniture in proportion. In some cases the important people are shown bigger than the servants (such as in the Bayuex Tapestry, Plate 1 where King Edward speaks to Harold).

As an example, one of the first chest I made was a six board chest based on Jewish merchants from the 'Cantigas de Santa Mari´a', El Escorial, Madrid, Spain See Figure 1. I used the measurement of my hand (12cm) to simulate the way one of the merchants is depicted holding the chest and because I could not examine or see all parts of the chests I had to make a number of assumptions about the chests:

1. I assumed the chests were square in cross section based on the end of the chest on the right of the picture;
2. I assumed that the brackets on either end of the chest on the chest on the left were corner brackets and were on all four corners of the chest;
3. I decided to make a one piece lid, even though the chest on the right could indicate that the chest lid was made with an inside edge, possibly with thin timbers nailed to the underneath of the lid to form a boxed lid;
4. For the thickness of timber I decide to use what timber was available, I had some pine planks so I used them. See Photo 7.

Of course you may end up with different measurements based on the size of your hand. Other body measurements could be used or even items in or around the chest.

Figure No. 1. Cantigas de Santa Mari´a (Canticles of Holy Mary). Reign of Alfonso X of Castile, "the Wise" (1221-1284). Jewish merchants, Library of El Escorial. Madrid. Spain, National Heritage.

Photo 7, replica chest based on an illumination of some Jewish merchants from the Cantigas de Santa Mari´a, made of pine.[10]

---

[10] Cantigas de Santa Mari´a (Canticles of Holy Mary). Reign of Alfonso X of Castile, "the Wise" (1221-1284). Jewish merchants Library of El Escorial. Madrid. Spain. National Heritage.

Marking Out and Cutting

NEVER assume the timber you buy from any supplier is the dimensions it is marked as or square, you must always check that the timber is true and square and make it so before you start marking out. Use a set square mark and then cut or shave/plane off until it is true and square.

Some suppliers will mark the length of the timber on one of the ends with a permanent marker so this will need to be cut off. Some suppliers us a sticky label for bar coding, this label should be removed as soon as possible to avoid making the wood with a photographic effect.

Buy more timber than you need, it will all get used at some stage.

"You can't stretch timber like Jesus did" (The Infancy Gospel of Thomas 13:1).

ALWAYS use pencil to mark out work and never use pen or permanent marker because this will permanently mar the work. *"Measure twice, cut once"* is an old saying but it is true, check your measurement before cutting and wasting your timber, effort and time. Ensure your pencil and saw are sharp and cut to the outside of the marked lines. Mark the pieces with a symbol denoting where the piece fits in a place which will not be seen when the work is complete (i.e. "End No. 1"). As far as it can be determined there is currently no evidence for marking pieces of Viking furniture to show how they are fitted together like the carpenter's marks on beams for a timber framed house.

Cutting order is important because for some projects the size and shape of the preceding pieces will affect the next piece (i.e. Oseberg Chest no. 178), and it helps to avoid un-necessary wastage.

Chest Fittings

Chest fittings consists of: hinges, hasps, hasp plates, corner brackets and locks. Chest can be locked with a padlock on a hasp, or a lock holding hasps in place. The size and complexity of locks increased with time, from spring, slide, turn key and combination.

The simplest of hinges is a hook and eye hinge (see Photo 8), where the back piece has an eye through which a hook on top of the lid of the chest swivels through. The eye of the back piece would have originally been hot punched but can be drilled. The hooks would have been beaten hot to a tapering long point but you can cut and grind down to shape, then bent into a tight curve, inserted in to the eye of the back piece and closed over to make a loop and a pivot point. See Table 5. for different chests that use hook and eye hinges.

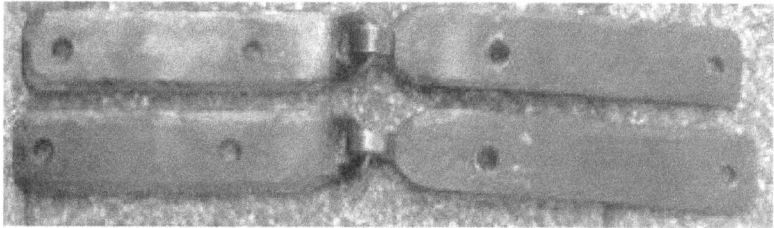

Photo 8. hook and eye hinges.

Table 5. Hook and eye hinge has been used on a range of chests.	
Cumwhitton chest	Cumwhitton in a Viking-age cemetery, Tullie House Museum and Art Gallery, Carlisle, Cumbria, in England, Grave No. 85.
Birka chest	Birka, Sweden. Grave no. Bj854
Hedeby chest	Hedeby Viking Museum, *Haithabu*, Schelswig-Holstein, Germany.
Lejre chest	Lejre (West of Roskilde, Denmark) Grave no. 1160.
Lejre casket	Lejre (West of Roskilde, Denmark) Grave no. 321.
Mästermyr chest	Statens Historiska Museum, Stockholm, No. 21592
Voxtorp Church chest	Voxtorp Church, Småland, Sweden, c. 1200. Statens Historiska Museum, Stockholm, No. 4094.

Method for making hook and eye hinge.

Blacksmithing:
1. Drift hole;
2. Beat and bend hook;
3. Join, shape to chest, hot punch nail holes.

Cold:
1. Drill eye hole;
2. Cut and grind hook;
3. Bend and join;
4. Shape to chest;
5. Drill nail holes.

Chest Locks

The wealthier you were, the greater the need to secure your valuables, hence, the need for lockable storages, such chests. Chest locks have increased in complexity with time and the cleverness of lock pickers. Chests could be secured by a padlock which secured a hasp in place, or a hasped lock with a bar that was slid (internally) by a key to secure the hasp in place and thus the lid that the hasp was attached to. Some larger chests were provided with multiple hasps, locks or a combination of the two. Early padlocks consisted of spring depressed by a key, once the spring was depressed the hasp of the padlock could be removed from the padlock, thus allowing the hasp to be opened and therefor the chest.

Early chest locks consisted of six types:
- Spring displacement, either parallel to the front of the chest (e.g. Mästermyr no. 4 or at 90 degrees to the front of the chest (i.e. Oseberg 178);
- Slide, the bigger the chest the greater number of hasps required to secure the slide;
    o Single hasp (i.e. Kaargarden);
    o Double hasps (i.e. Winchester lock) and Mästermyr, see photo 9;
    o Triple hasps (i.e. Osberg 149);
    o Slide and knob (i.e. Frykat);
- And padlocks;
    o Barrel (Single spring) (Coppergate York). Photo 10;
- Cowbell (Multiple springs) (i.e. Mästermyr).

Photo 9. Stephen (Sven) Wyley with his Mästermyr tool box with double hasp lock I made with help from Gaz Bannister and Driffa and John Russell. The chest was made by Natalie Vasilaka.

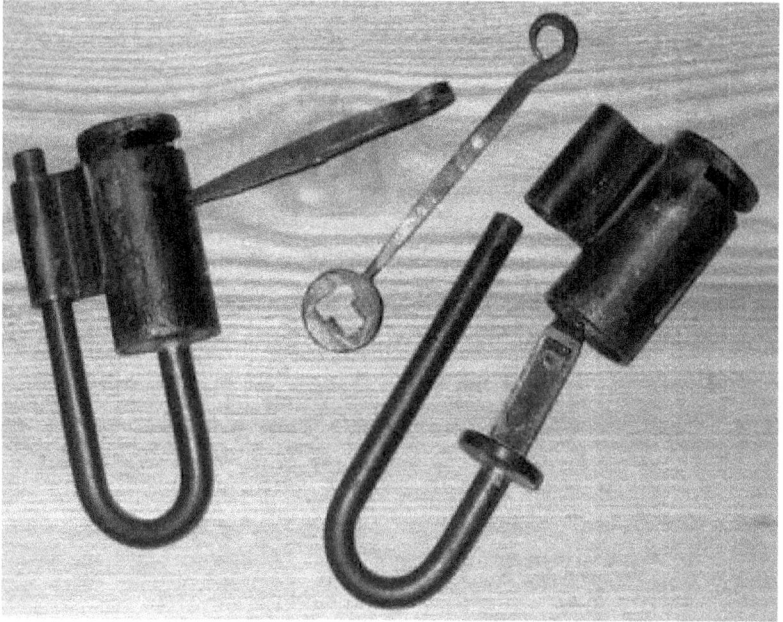

Photo 10. Barrel padlocks, left key in slot of a padlock, the spare key and a padlock opened.

Construction Projects

The Tool

The Coppergate Mallet.

The Coppergate mallet	16 – 22 Coppergate, York, England	When: 930 – 975 A.D.
Stored: Yorkshire Museum, York, United Kingdom	Collection no: 8186	Material: Willow head with hazel handle
Size:	Size of handle - 175mm * 18mm round	Size of head - 118mm *36mm

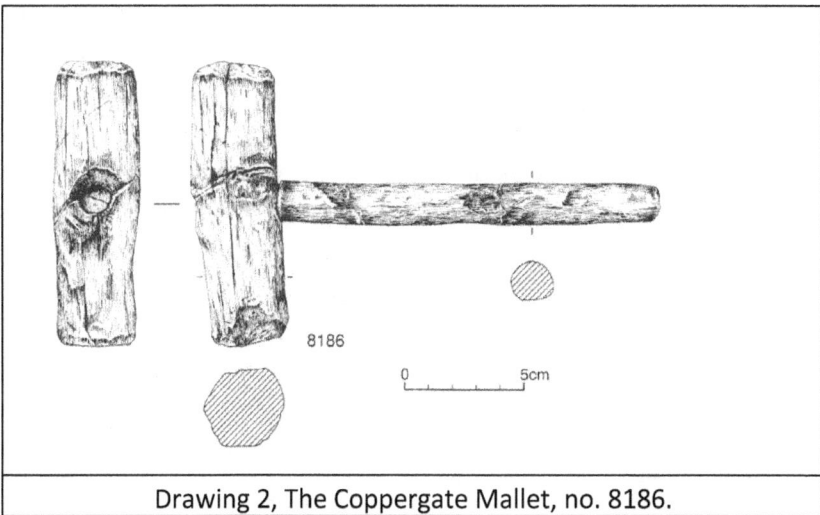

Drawing 2, The Coppergate Mallet, no. 8186.

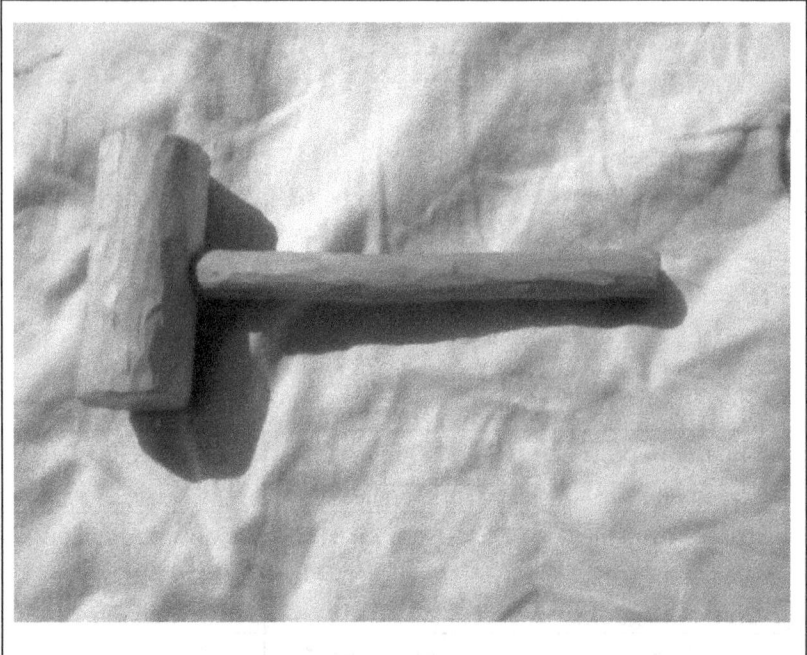

Photo 11. A mallet made with a birch head and plum handle.

A wooden mallet is a utilitarian tool made simply for thwacking things and easily replaced if broken or lost but as a first project it is one of the easiest to make and it will come in handy for all of the other projects in this volume and those to come. The Coppergate mallet is contemporary to that found in the Oseberg boat burial and the four found in Dublin[11], and is similar to those found in the 16th century ship the Mary Rose, see Table 6.

---

[11] Morris C.A. 1984a, fig. 165 W139i-iv (unpublished) Anglo-Saxon and Medieval Woodworking Crafts: the Manufacture and Use of Domestic and Utilitarian Wooden Artefacts in the British Isles, 400-1500 A.D., Ph.D. thesis, University of Cambridge.

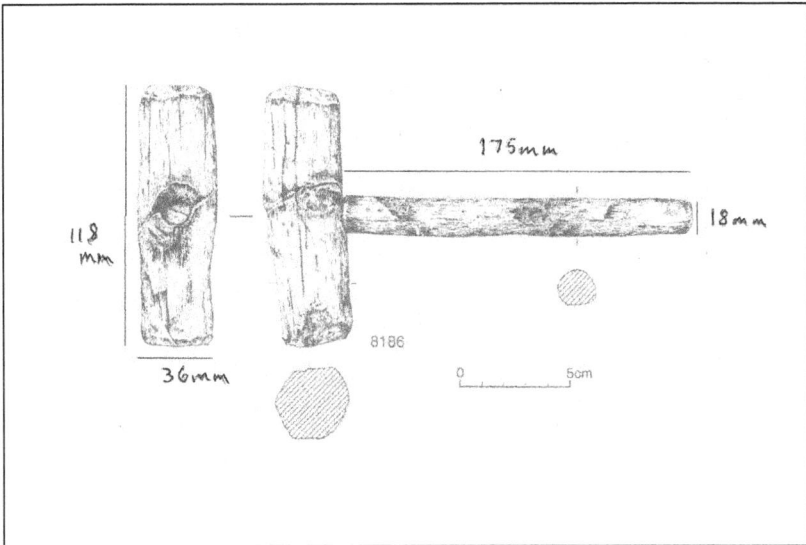

Drawing 3. The mallet with measurements

Table 6, Mallet comparison.				
Find details	Time period	Size of handle	Size of head	Timber
Oseberg (integral handle) flax beater	800-850 A.D.	300mm * 50mm near head to 8mm at end.	360mm *200mm	Unknown
Novgorod (integral handle)[12]	9th to 10th Century	30cm * 5cm near head to 8cm at end.	25cm *16cm at top, 18 cm near handle	Unknown

[12] Gorodishche, wooden objects, line drawing only, No. 8, figure 11.14, page 32, The Archaeology of Novgorod, Russia, Lincoln 1992. Measurements based on drawing 1mm = 1cm.

York, from 16-22 Coppergate, no.8186 (is a block on a sepa-rate wooden handle)	930 – 975 A.D.	175mm * 18mm round	118mm *36mm	Willow head with hazel handle
Mary Rose (81A5699 (is a hexagonal block on a separate wooden handle)	Tudor 1545 A.D.	250mm* 25mm	220mm *80mm	Oak [13]

Materials.
- Head - Tree branch (easiest to work green), 118mm in length, and 36mm in diameter.
- Handle - Tree branch (easiest to work green) or suitable dowel, 175mm in length, and 18mm in diameter. Note: if you use green wood for the handle, it must be allowed to dry (and hence finish shrinking) before mounting on head.

Tools.
Ruler, pencil, saw, mallet, draw knife, shaving horse or bench vice.

Construction Instructions. See Drawing 3 and Photo 11:
1. Remove bark from tree branch with draw knife, a bench vice or shaving horse can be used to secure the tree branch;
2. Mark out and cut head to length with saw;
3. Mark out and cut handle to length with saw;
4. Drill hole through head for handle;
5. Cut a 2mm wide cut in the end of handle;

---

[13] Others woods used in other mallets include ash, elm and maple.

6. Insert handle into head, insert wooden wedge into end of handle, secure the handle then hit home with mallet.

The Stool

The three legged Lund stool

The Lund stool	Lund (KV Färgaren 22), Sweden.	When: First half of the 11th century
Stored: Kulturen, Lund, Sweden	Collection no: KM53.426:1074	Material: Wood – Birch.
Top only - size: 395mm* 240mm * 20-25mm		

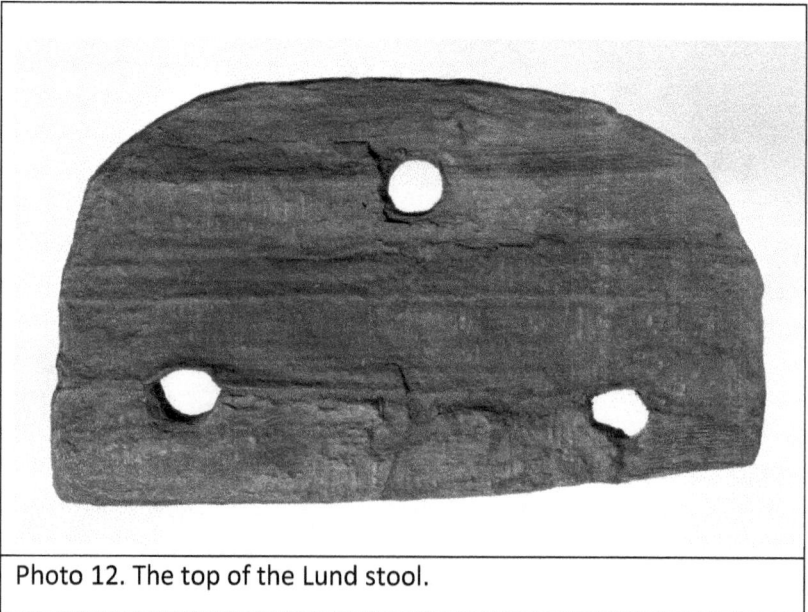

Photo 12. The top of the Lund stool.

Photo 13. A completed Lund stool, the top was hewn from a log and the legs were made from tree limbs.

Three legged stools[14] have been a piece of furniture that has been around for a long time. The earliest extant find is from Egypt dating from the New Kingdom (c. 1567 – 1085 B.C.)[15]. Stools varied in size and shape, with the purpose of keeping your bottom off the ground while you went about your domestic or manufacturing activities. See Table 7 for a comparison of some of the extant finds.

---

[14] See the Hedeby stool for a 4 legged version. "[dating from] *AD 850-1066 Oak, 29.6 x 20.6cm and 2.1cm thick. Leg holes are 2.6cm in diameter and the oak legs where wedged. The complete height was 19.5cm*", from <u>Westphal</u> (2006).
[15] From Egyptian Wood Working and Furniture, p. 46, Picture 54 '(National Museum of Scotland, Edinburgh, 1956.107.'

This project is based on the "*Lund*" Viking stool which dates from the first half of the 11th century (Kulturen, Lund, Sweden, KM53. 426), see Photo12. The remains of the stool only consists of the birch seat which forms the rough shape of a "*D*", 40cm across the straight edge, and is perforated by three holes for the legs. Two of the leg holes are situated in opposite corners of the straight side, while the third hole is set in the middle of the curved side. The three holes form a triangle. The front (straight) edge slopes towards one side, it is not discernible if this is the original shape or later damage. I have assumed it is damage and the plans provide for a straight front.[16]

Table 7. Three legged Stools – Comparison of stools. [17]				
Place	Date	Collection	Material	Dimensions
Lund, Sweden	1000-1100, most probably 11th	KM 66166:1924	Oak	380mm*240mm * 20-25mm
Sweden	1000-1100	KM 57135:561	Oak	440mm*220mm * 20-25mm
Lund, Sweden	First half of the 11th century	Lund (KV Färgaren 22), Sweden. – KM 53.436:1074	Birch	395mm* 240mm * 20-25mm
York, 16-22	Mid 14th to	No. 8948	Oak – tangentially split board	Length 56.5cm, Width 38.5

---

[16] Other similar stools were found in 16-22 Coppergate (York, No. 8946, 8947 and 8948), Winchester (3441), Fishamble Street Dublin (M338), and Sweden (KM 66166:1924) and (KM 57135:561).

[17] Thanks to Gunilla Gardelin (Cultural Environment Developer / Curator - Cultural Historical Association of Southern Sweden) for the thickness measurement and other information.

Cop-pergate	15th century			cm, thickness 5.6cm. Leg holes 36mm in diameter.

Materials.
- The Seat, timber plank, 40cm*20cm*3cm, from experience I recommend using a plank at least 32mm thick, 20mm planks tend split across the seat where the leg holes are located.
- The 3 Legs, timber dowel or tree limbs, 'X' cm *35mm or suitable tree limbs, were "X" is the distance from the ground to the back of your knees in the seated position.
- wood glue, which can be omitted if you make the legs demountable, which makes easier to flat pack for travel.
- and linseed oil.

Notes:
1) I am relatively tall and I use legs 45cm long but I recommend a maximum of 50cm for reasons of stability.
2) I recommend making a plan in cardboard or in wood to use as a template for later use.

Tools.

Pencil and ruler, jigsaw (electrical) or coping saw, hole cutter or spade bit on a drill, hand saw, rasp or knife.

Construction Instructions.

1. Select a suitable piece of timber for the seat. Use Birch if possible or some equivalent wood. If you have the time, skill and tools you can split and trim down a plank from a log.

2. Mark out timber as per plan using a pencil. Make sure to avoid cutting through knots. See Drawing no. 4.

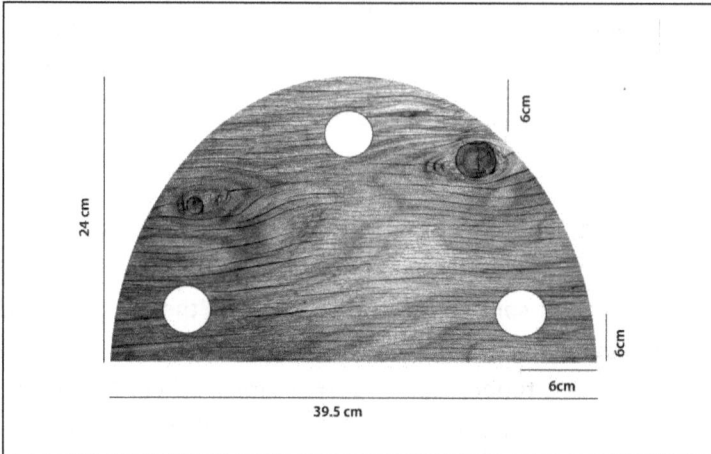

Drawing 4. The size of timber you need and hole placement. Drawing by Jenny Baker.

Drawing 5. Looking at front of stool, showing angle of front legs.

3. Cut timber to shape using coping saw or jigsaw.

4. Cut leg holes 32mm wide in the appropriate places on the seat using a hole saw or spade bit and a jib, see Photos 14 &15. Angle front

two holes at 10° from the horizontal plane of the seat using a protractor or jig to form the base of a three dimensional prism. See drawing 5 & 6.

Photo 14, Drill, hole saw and jig used to cut angled leg holes. Note also the waste timber under the stool top, which also protects the bench top from damage.

Photo 15, The first leg hole drilled in the stool top, from the top.

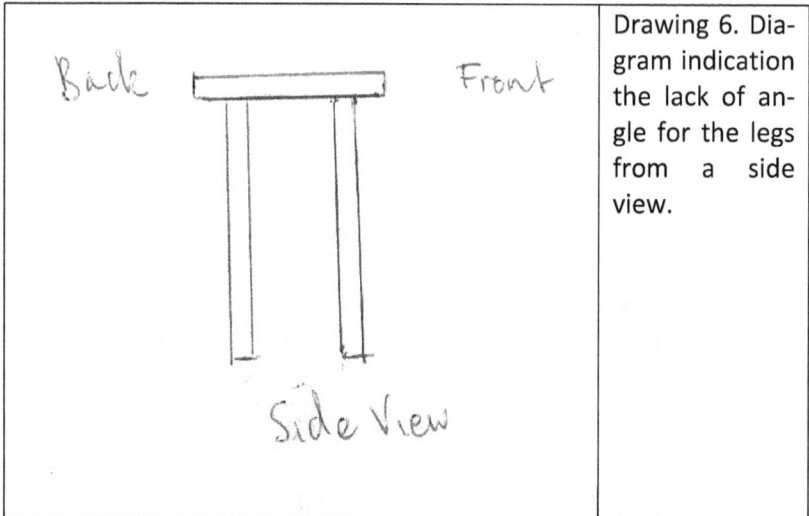

Drawing 6. Diagram indication the lack of angle for the legs from a side view.

5. Cut three legs to the require lengths, cutting the ends at a 10° angle at top and bottom to match the seat and the ground. See Drawings 5 & 6 and Photos 14 & 15.

6. Carve the tops of the legs to fit into the holes in the seat using a drawknife. The legs must be a snug fit or they will make the stool unstable. Cut off excess leg if poking up through the hole with a saw.

7. If you are going to make permanent joints it is recommended that a slot is cut down the centre of the end of the top of the leg, inserting a wood wedge into the saw cut with the leg in place in the leg hole, with the liberal application of glue. Clean off the excess glue and allow to dry.

8. Sand edges and upper surface.

9. Apply linseed oil or danish oil to finish stool.

10. Write or carve your name, and the year it was made and group on bottom of seat. It helps getting your stool back when it gets lost.

Recommendations:

1)   Timber will dry out over time, for a good leg fit soak the top ends of the legs in water prior to use.

2) An annual coating of suitable oil (i.e. linseed oil) for the preservation of the wood. There is nothing more annoying at an event than when your legs fall out.

3) Keep a piece of dowel cut to the 10 degree angle as a guide when cutting other legs to the same angle.

The Chest

The Oseberg 178 Chest

The Oseberg Chest	Found: Oseberg farm, Tonsberg, Vestfold, Norway	Dating: 800 – 850 A.D
Stored: Viking Ship Museum, Oslo, Norway.	Collection no: 178	Material: Pine and iron.
Size: 665 * 210 * 310mm, thickness - unknown		

Photo 16. The original Oseberg chest no. 178.

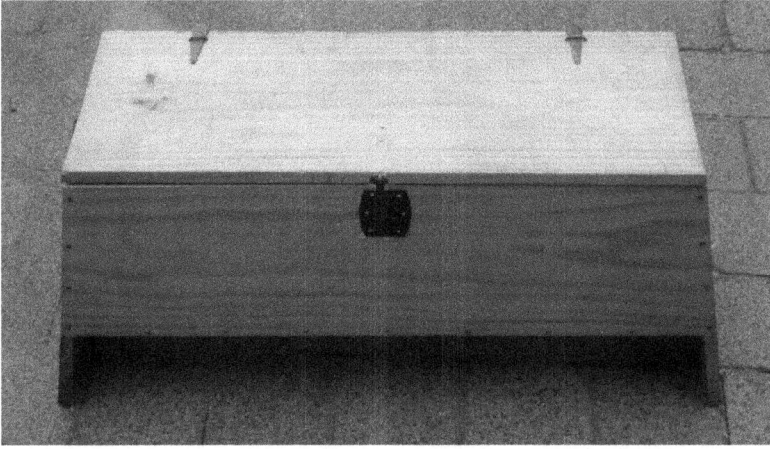

Photo 17 Oseberg chest 178 with lock.

The Oseberg Viking ship burial (800 – 850 A.D.) is an extensive find including the ship, a sled, a cart and multiple chests, including no. 178, which was similar to the others in the burial consisting of a 6 boarded trapezoidal prism made from oak (for other Trapezoidal chests see table 8). The bottom was mortised into the ends with halving joint across the ends, the front and back fitting to the ends with a cut out join, and the lid is flat, see drawing 9. The hinges consist of a number of staples and a hook, producing a pivot point, see drawing 10. The lock is a spring lock at 90 degrees to the front of the chest behind a lock plate, the key enters a key hole in the lock plate, rotates and depresses the spring, the lid and the spring can now be lifted upwards and open, locking is just a reverse process not requiring the key, see drawing 12.

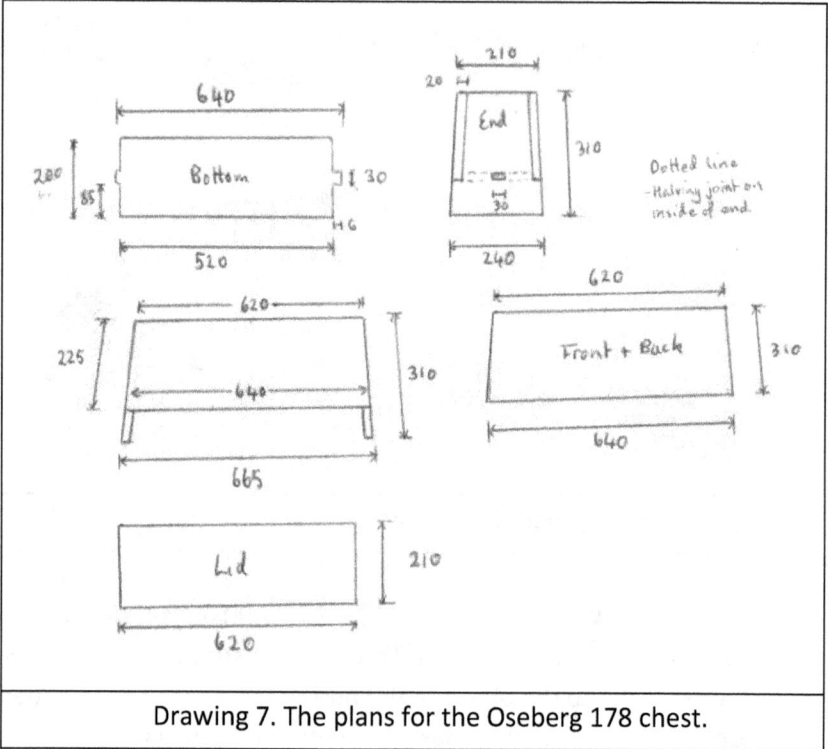

Drawing 7. The plans for the Oseberg 178 chest.

Table 8. Trapezoid chest comparison.				
Place	Date	Collection	Material	Dimensions
Oseberg farm, Tonsberg, Vestfold, Norway	800 – 850 A.D	Viking Ship Museum, Oslo, Norway. No. 149	Oak and iron	L (base 156cm, top 104cm), W (base 36cm, top 28cm), H 41cm.
Oseberg farm, Tonsberg, Vestfold, Norway	800 – 850 A.D	Viking Ship Museum, Oslo, Norway. No. 156.	Oak and iron	L (base 113cm, top 108cm), W (base 29cm, top 32cm), H 35cm.
Oseberg farm, Tonsberg, Vestfold, Norway	800 – 850 A.D	Viking Ship Museum, Oslo, Norway. No. 178.	Oak and iron.	L (base 66.5cm, top 52cm), W (base 24cm, top 21cm), H 31cm.
Birka Island, Sweden	9th century	Grave 854, Birka Museum, Björkö	Only the metal fitting are extant, speculations that it had a	There is a copy displayed in the museum

				trapezoid shape.	
Hedeby Harbour, Germany	Pre-982 A.D.	Viking Museum Haithabu	Oak and iron	L (base 52cm, top 42cm), W (base 23cm, top 22cm), H 27cm.	
Mästermyr, Sproge parish, Gotland, Sweden	1000 A.D.	Staten Historiska Museum, Stockholm, Sweden, Ref. 21592	Oak and iron	L (base 92cm, top 88cm), W (base 25.6cm, top 24cm), H 24.6cm.	
Townfoot Farm, Cumwhitton, Cumbria, England	10th Century	Grave 85, Tulee House Museum, England.	Acer sp. (Maple) and iron	L (base 46cm, top 38cm), W (base 16cm, top 11cm), H 20cm.	

Materials.

Timber (oak preferably but Pinus radiata will do) 28.5cm wide * 3.2 metres long. I recommend getting at least 4 metres of the timber just in case. See Table 9 for the timber cutting list.

Table 9. Showing cutting list for an original Oseberg 178 chest.	
Bottom	64.5cm
End * 2	31cm
Front	66.5cm
Back	66.5cm
Lid	62cm

Note: These measurements are based on the original chest but plans are available for a larger version, see appendix 1.

Nails (*26, 30 just in case), glue, mild steel for fittings (hinges, hasp, lock plate and key), spring steel for the spring in the lock, see Table 10.

Table 10. Metal requirements.			
Steel	Hinges	Lock	Keys
Mild Flat bar (3mm)	300mm	90*10mm	2 *150*5mm
Mild sheet (2mm)	X	80*70mm	X
Wire (2mm)	810mm	30mm for ring	X
Spring steel	X	60mm*10mm	X

Note: Making 2 keys is recommend just in case you loose one. And don't store the spare key in the chest. Don't laugh it has happened.

Tools.

Rulers (30cm, 60cm, and 100cm), set square, chisels (25mm & 7mm, plane edge), files (course - flat bastard, fine - square bastard), tenon saw, hand saw, plane, 'G' clamps, scrap wood (for holding work without damaging it), claw hammer, wooden mallet, bench vice, pencil, hand drill (with 4mm bit).

Construction Instructions.

Bottom.

1. Mark out bottom with pencil, set square and ruler.

2. Cut the angled ends on either end first bottom.

    a. Place each end of the bottom in a bench vice and cut the end cuts on either side of the tenon, angle the cuts to match the angle of the end of the bottom. See Photo 18.

    b. Turn the bottom on its side in the vice and cut down at an angle to meet the end cut on one side of the end. See Photo 19.

    c. Check angle and modify if required.

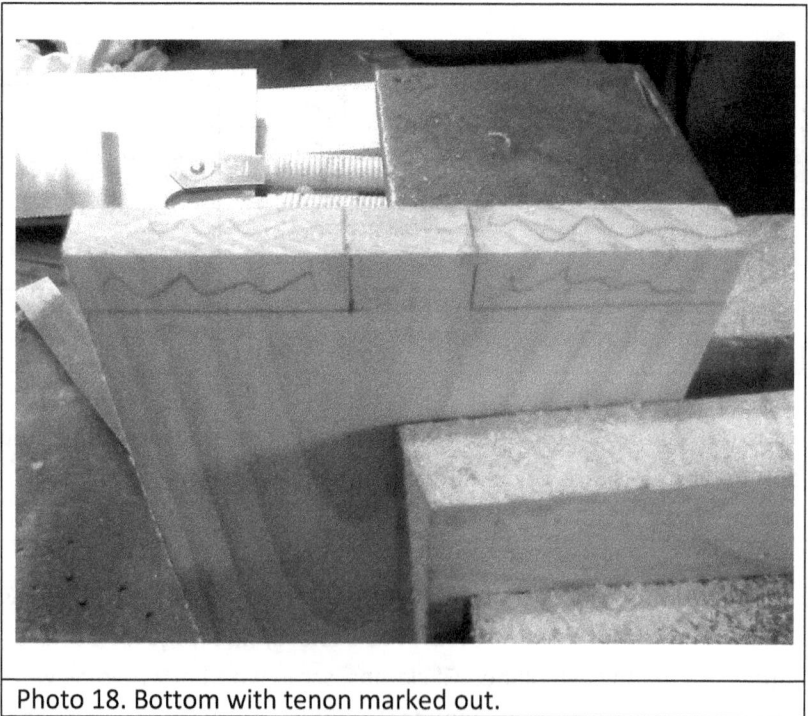

Photo 18. Bottom with tenon marked out.

Photo 19. Bottom with tenons cut.

End.

   3. Mark out ends with pencil, set square and ruler.

   4. Cut the ends

      a. Cut the inserts for the front and back on the sides of the ends. See Photo 20.

      b. Turn the end on its side and cut downwards at the angle to match the ends.

      c. Turn the end on the bottom end and cut the side inserts.

Photo 20. End with joints cut out on either side.

Halving joint in ends.

    5. Mark out halving joint between both ends of the side cut outs.

        a. clamp end to bench, exposing the halving joint

        b. cut halving joint on either side with a tenon saw, down to 8mm depth (20mm wood). See Photo 21.

        c. chisel out halving joint. See Photo 22.

        d. clean up joint.

Photo 21. End, saw cuts with tenon saw on either side of the halving joint.

Photo 22. End, halving joint chiseled out.

Mortice in ends.

6. Mark out the mortice.

a. using a sharp knife (Stanley/ box cutter) to make the outside of the mortice.

b. clamp the end to the bench.

c. chisel or drill out mortice from either side. It is important to have a clean tenon on the outside because people will see the error. See Photo 23.

7. Fit ends to the bottom.

a. this may entail modify the tenon or the mortice to fit snugly with rasp and file. See Photos 24& 25.

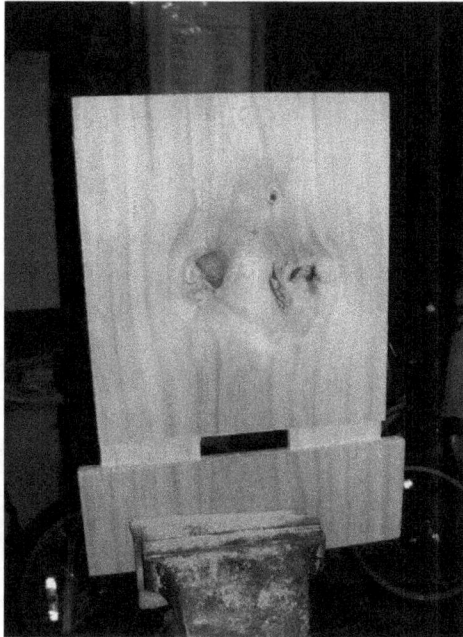

Photo 23. End, Mortise cut through halving joint

Photo 24. The Bottom is held in the bench vice while the end is fitted to the tenon of the bottom.

Photo 25. The end fits the mortice.

Front and back.

8. Use the fitted bottom and ends to mark out the front and back.

   a. compare the angle of each end of the chest and modify the joints until they match. See Photos 26 and 27.

   b. cut out front and attach with nails (skew nails for better security) and glue. 3 nails per side and 6 nails along bottom is recommended.

   c. repeat for back of chest. See Photos 28 & 29.

Photo 26. Marking and ensuring the angle of the front and back are the same. There may be some play in the joint which will be able to be used to obtain the same angle on each end.

Photo 27. Marking and ensuring the angle of the front and back are the same.

Photo 28. Placement of nails compared to the halving joint.

Photo 29. Placement and skewing of the nails.

Photo 30. The box of the chest.

Lid.

    9. Place the box upside down onto the timber for the lid.

        a. mark out box out line on timber using a pencil.

        b. cut the angled ends to match the angled ends of the chest.

        c. cut to width. See Photos 30 & 31.

    10. Plane and scrape to clean up chest. Apply oil as a preservative.

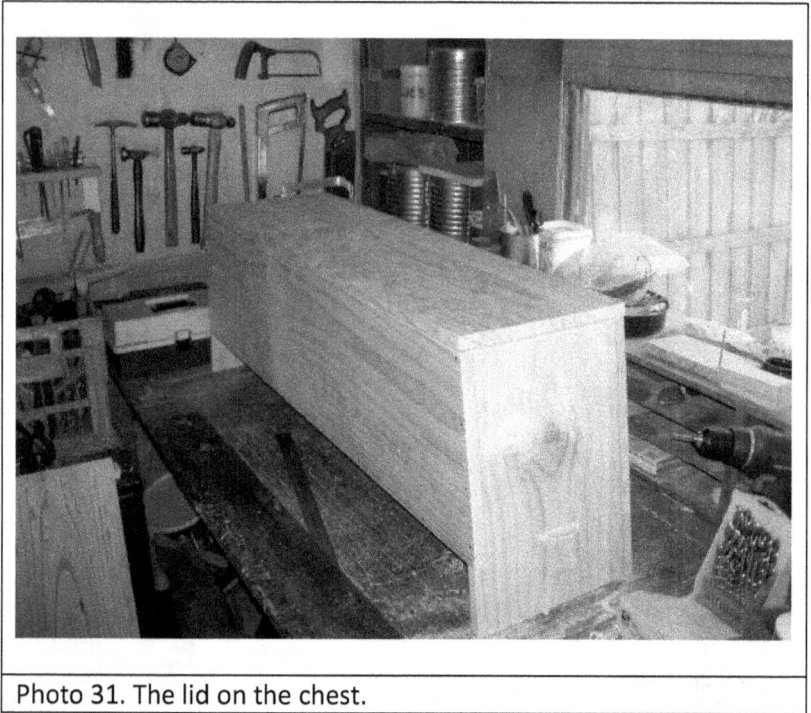

Photo 31. The lid on the chest.

Notes:

1) Cut each piece in sequence based on the piece that came before it, they may differ from your plans.

2) Boiled linseed oil (1 part boiled linseed oil to 3 parts mineral turpentine) is recommended for preserving the timber.

Making the Hinges.

Top of hinge.

11. Mark out hinge top, see Drawing 8.

12. Cut out top pieces from flat bar. Grind to shape.

13. Bend back end of hinge in a tight curve so as to enter a hole in the back edge of the lid,

14. Bend the front end of the hinge roughly at 90 degrees. See Drawing 9 and Photos 32 & 33.

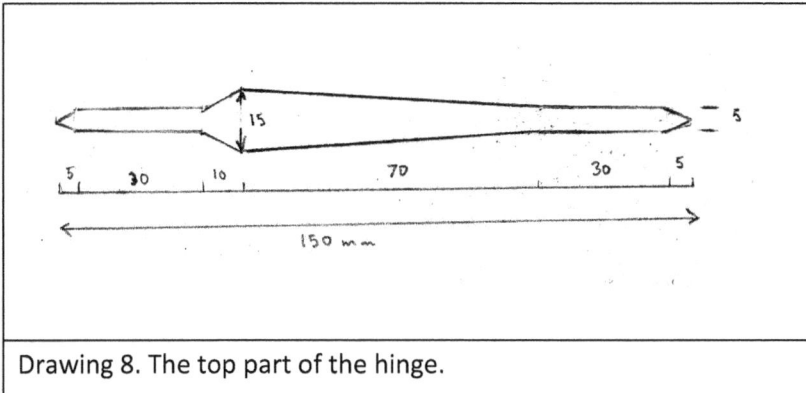

Drawing 8. The top part of the hinge.

Staples.

15. Cut and shape wire staples (two long loop staples and 4 short staples). Ensure ends of staples have points for easily entering timber.

16. Bend the last 3 cm of the long staples at right angles to go into the chest.

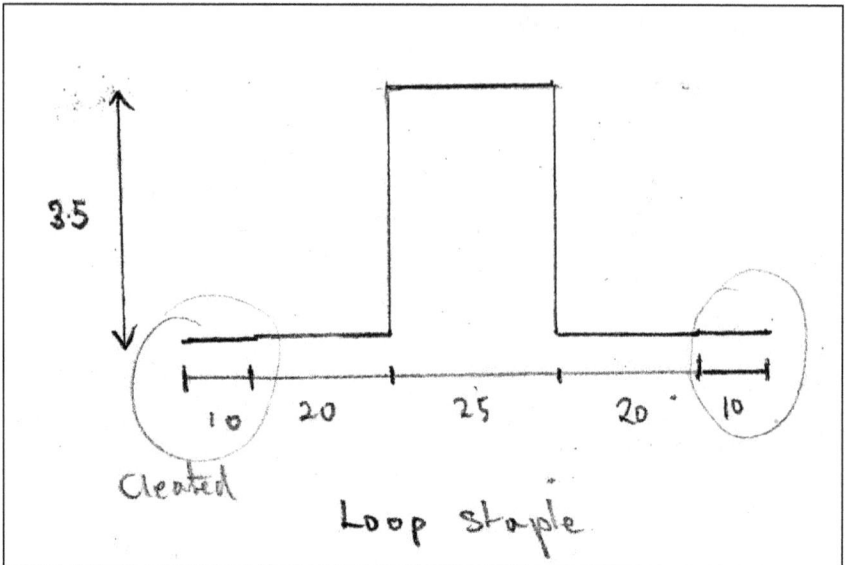

3.5

Cleated

10   20   25   20   10

Loop staple

Drawing no. 9. The Loop staple which formed the bottom part of the pivot for the hinges.

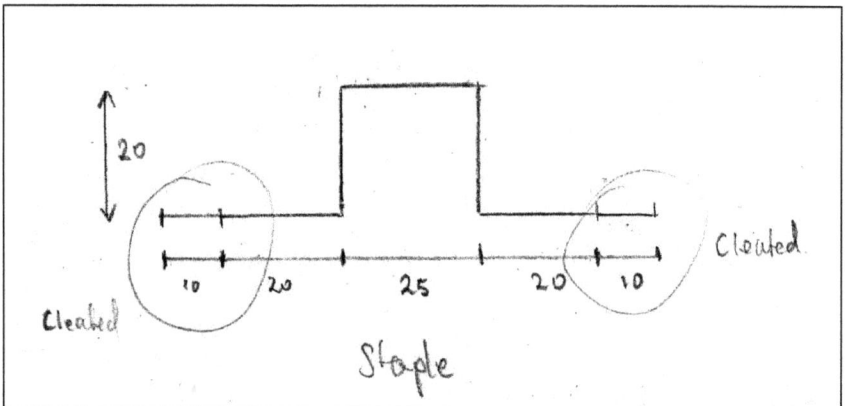

20

Cleated

Cleated

10   20   25   20   10

Staple

Drawing no. 10. The staple which held the top part of the hinge in place and the Loop staple in place.

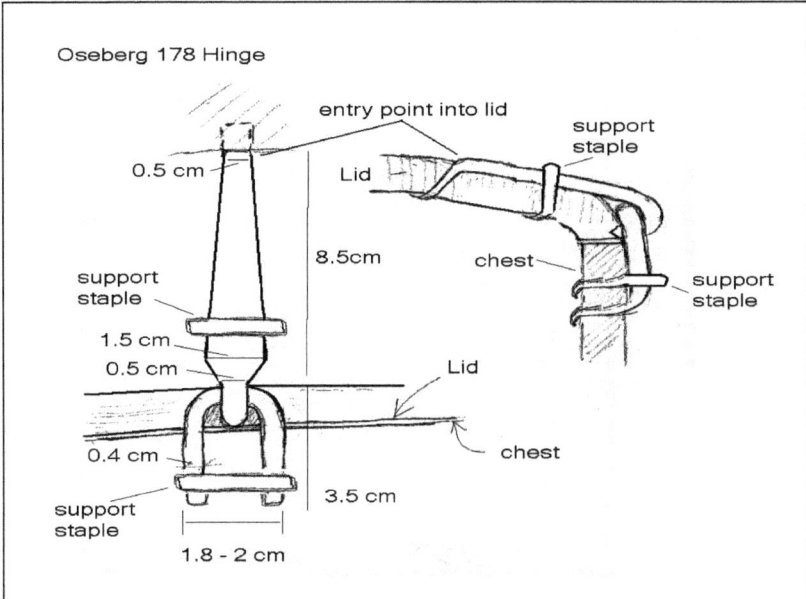

Oseberg 178 Hinge

entry point into lid

support staple

0.5 cm

Lid

8.5cm

chest

support staple

support staple

1.5 cm

0.5 cm

Lid

0.4 cm

chest

support staple

3.5 cm

1.8 - 2 cm

Drawing 11. One of the hinges consisting of a top hook, a back "U" and two support staples.

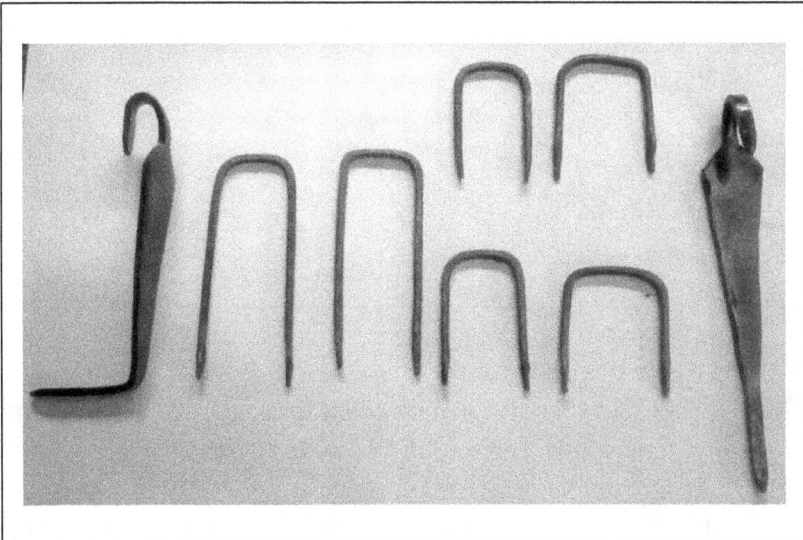

Photo 32. The hinges and staples.

Adding fittings.

Top part of hinge. See Photo 33 & 34.

1. Drill hole for back part of hinge into the centre of the back edge of the lid 14 cm from end of chest, about 2cm deep and wide enough to take hook.
2. Fit hook in to hole in back of lid and drill hole through top of lid for front part of hinge where the front part of the hinge touches the lid, and 14 cm from end of chest.
3. Insert into back end of hinge into hole in back edge of the lid.
4. Insert front of hinge into top of chest lid.
5. Hammer front of hinge down into top of lid.

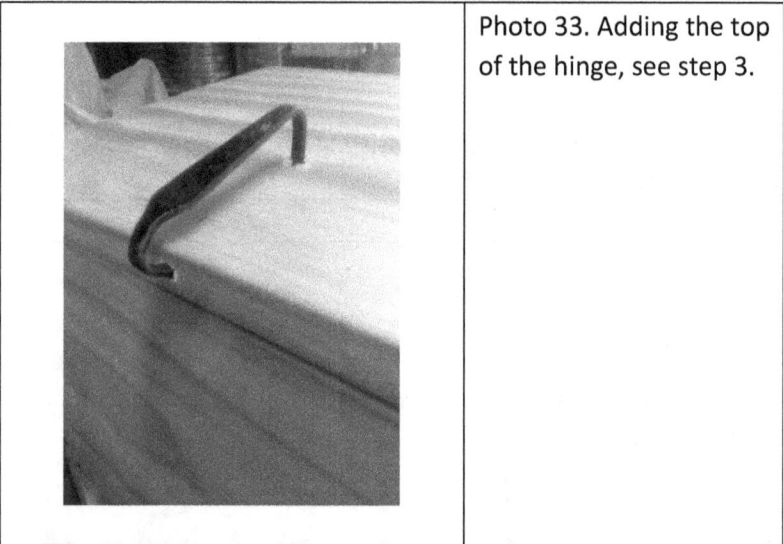

Photo 33. Adding the top of the hinge, see step 3.

Loop staple (the pivot for the hinge).

- ensure there is full movement of hinge with placement of loop staple. There will be some play in the hinges.

6. Drill holes for staple connecting the top of hinge to back of lid.
   a) insert staple through back of hinge in back edge of lid.
   b) And hammer into place.

Photo 34. Hinge loop in the back of the lid and held in place by the Loop staple. See step 6.

Back Staple.

7. Drill holes for staple on the outside of the Loop Staple about 2cm from the top of the back of chest.
   a) insert staple through back of chest.
   b) Hammer into place.

Top staple.

8. Drill holes for top staple about 2.5cm from back end of lid on either side of top part of hinge.
   a) insert staple over top of hinge on top of lid and hammer into place.

|  | Photo 35. Hinge with all 3 staples. |

Cleat hinge and staples in place.

9. Turn chest turtle, place anvil under hasp and staples, bend over ends of hasp and staples cleating them into inside of lid and back of chest. See Photo 36.

Photo 36. The inside of the lid and back of the chest showing the cleated parts of the hinge and the staples, cleated in opposite directions to ensure that they can easily be pulled out.

Making the lock

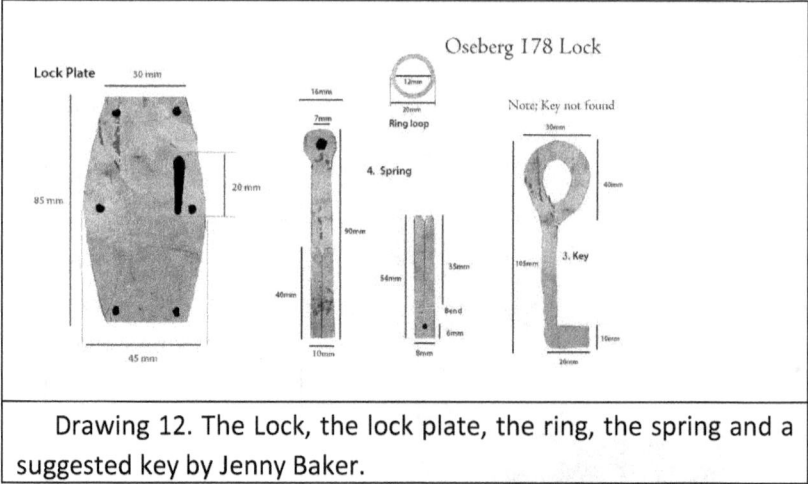

Drawing 12. The Lock, the lock plate, the ring, the spring and a suggested key by Jenny Baker.

Photo 37. The spring, the lock plate and the key. [18]

---

[18] There was no key found with the chest. A suggested key shape is provided based on the key(s) found in Paviken (Västergarn parish, on the island of Gotland, Sweden, during Per Lundström's 1968-1971 excavations, or the Birka grave finds of Sweden (Graves number 270 and 585).

1. The Lock plate
   a) Mark out lock plate out from sheet.
   b) Cut out the lock plate using cold chisel, hammer and anvil, or an electrical device (jigsaw with metal cutting blade), an angle grinder with a cutting disc. Note: wear the PPE (safety glasses and hearing protection).
2. The Spring
   a) Mark out and cut out the bar (mild steel), grind to shape and drill hole in rounded end.
   b) Mark out and cut out the spring (spring steel), grind to shape, cold punch or drill hole in one end (you may need to anneal the end of the spring to enable the drilling), cut slit in other end of spring.
   c) Bend end of spring to form a 'tick' shape using a butane torch or forge.
   d) Rivet spring to base of bar.
3. The Key
   a) Mark out of mild steel bar, use forge or gas torch to heat up metal to bend ring in to shape, flatten end and bend bottom end of key as per Drawing 12. Ensuring the key will fit in the key hole.

Attaching the Lock and Hasp.
   a) Mark out centre of lid and front of chest.
   b) Chisel out hole in the front of the chest to take spring and provide access to the spring for the key. See Drawing 13.
   Note: Careful not to chisel through the back of the cavity.
   c) Mark centre of front of lid.
   d) Cut/chisel out front edge of lid so spring hangs down into the hole and engages the stop/cut out in the front of the chest behind the lock plate.
   e) Drill a holes in front edge of lid for a staple to hold the spring ring in place.

f) Hook ring on spring onto staple and hammer staple in place.

g) Place lock plate in place, drill nail holes based on holes in lock plate. See Photo 38.

h) Nail lock plate nails in into place.

i) Cleat nails over inside chest, cleat each nail in opposite directions to each other.

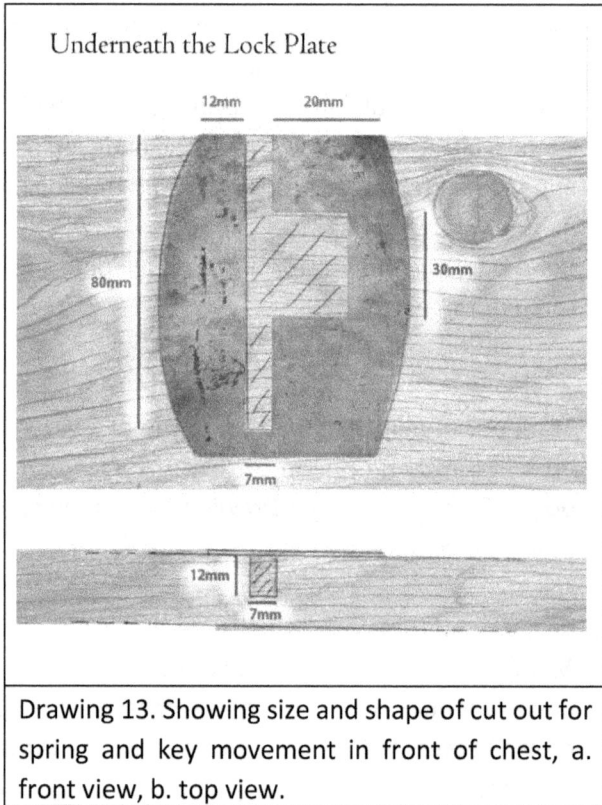

Underneath the Lock Plate

12mm   20mm

80mm   30mm

7mm

12mm

7mm

Drawing 13. Showing size and shape of cut out for spring and key movement in front of chest, a. front view, b. top view.

Photo 38. The lock in place on the chest.

The Bed

The Oseberg No. 2 bed

The Oseberg bed	Found: Oseberg farm, Tonsberg, Vestfold, Norway	Dating: 800 – 850 A.D
Stored: Viking Ship Museum, Oslo, Norway.	Collection no: 2	Material: Timber unknown
Size: 170cm long. 100cm wide and 75 cm high at the top of the side (and 100 cm at the top of a leg).		

Photo 39. The museum reproduction of the bed.

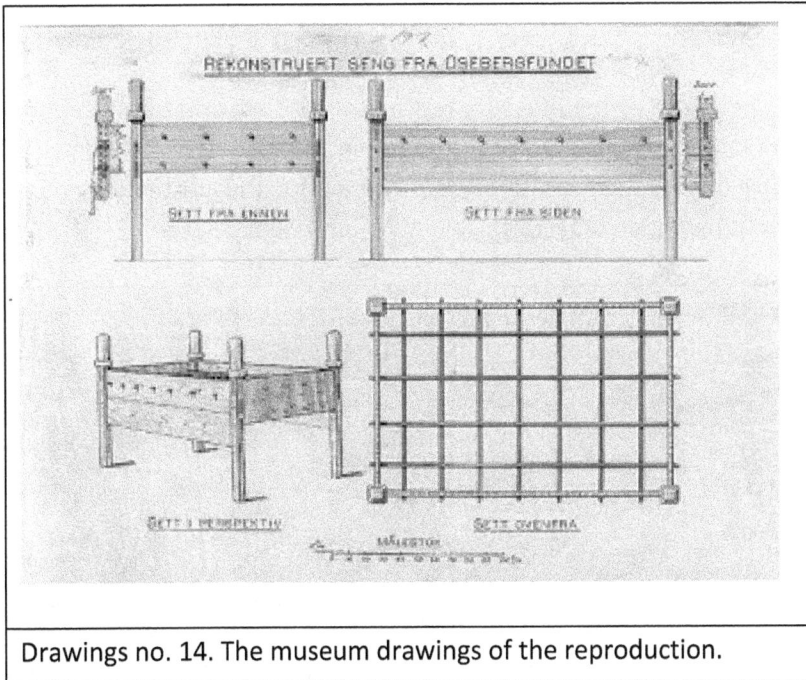

Drawings no. 14. The museum drawings of the reproduction.

Beds have been around for thousands of years, there are extant beds from the Egyptians (first Dynasty) consisting of a wooden frame with mortised and tenon, joints. It would appear from where the remains were found in higher status burials across a wide range of cultures and time span that it could be assumed that only the elite or wealthy of a society had beds.

The Oseberg bed no. 2 is just one of the beds found in the Oseberg ship burial (800 – 850 AD). The size of the original, based on the sizes of the replica made and the drawings is 170cm long. 100cm wide and 75 cm high at the top of the side (and 100 cm at the top of a leg). The replica for museum shows that the bed is held together with mortise and tenon joins and pegs, I have went without the doweling for ease of dismantling for use at re-enactment events.

Given the required bed size varies from re-enactor to re-enactor and must account for use at home or camping as well as the size of mattress being utilised, the method I provide below can be used for varied length and widths, see Appendix 3 for a version that is 2 metres long.

The photo of the museum replica shows dowel matress supports, with long axis dowel woven through the short axis dowe. As an alternative to dowel I have also provided details for a suggested roping pattern but wooden bracers and sheets of ply could be used.

Photo 40. The longer version of the bed in pine, sans rope support for mattress.

Table 11. Bed comparison.				
Place	Date	Collection	Material	Dimensions
Oseberg farm, Tonsberg, Vestfold, Norway (Horse Head)	800 – 850 CE	Universitetets Oldsaksamling, Oslo, Norway. No. 1 (K.69).[19]	Beech	L 175cm, W 184, H (Foot, 92cm *14cm), H (Horses head, 168cm *45cm).
Oseberg farm, Tonsberg, Vestfold, Norway (Plain)	800 – 850 CE	Universitetets Oldsaksamling, Oslo, Norway. No. 2	Timber	170cm long. 100cm wide and 75 cm high at the top of the side (and 100 cm at the top of a leg).
Gokstad Sandefjord, Vestfold, Norway	890 CE	Universitetets Oldsaksamling, Oslo, C10408.	Oak	Height 140cm at the animal heads.
Gokstad Sandefjord, Vestfold, Norway (plain)	890 CE	Universitetets Oldsaksamling, Oslo.	Unknown	Corner post 70-75cm. Endboards 1.09m * 28-30cm. Side board 2.27m * 28-30cm.

---

[19] '*Viking to Crusader*' provides a collection no. of K69 for the animal headed bed.

Materials.

Table 12. Material details.		
Beech, but Pine will do.	The size of the pieces.	The total length required.
2 Sides	170cm * 40cm * 2cm	340cm
2 ends	100cm * 24cm * 2cm	200cm
4 legs	100 * 7cm * 7cm	400cm
Rope	18 metres of at least 8 mm rope. Recommended fibres are linen or hemp.	

Tools.

Rulers (30cm, 60cm, and 100cm), Set square, chisels (25mm & 7mm, plane edge), files (course - flat bastard, fine - square bastard), tenon saw, hand saw, plane, 'G' clamps, scrap wood (for holding work without damaging it), hold downs, claw hammer, wooden mallet, work bench, bench vice, pencil, drill (with 4mm bit), spade drill bit (20mm).

Construction Instructions.

Sides and ends.

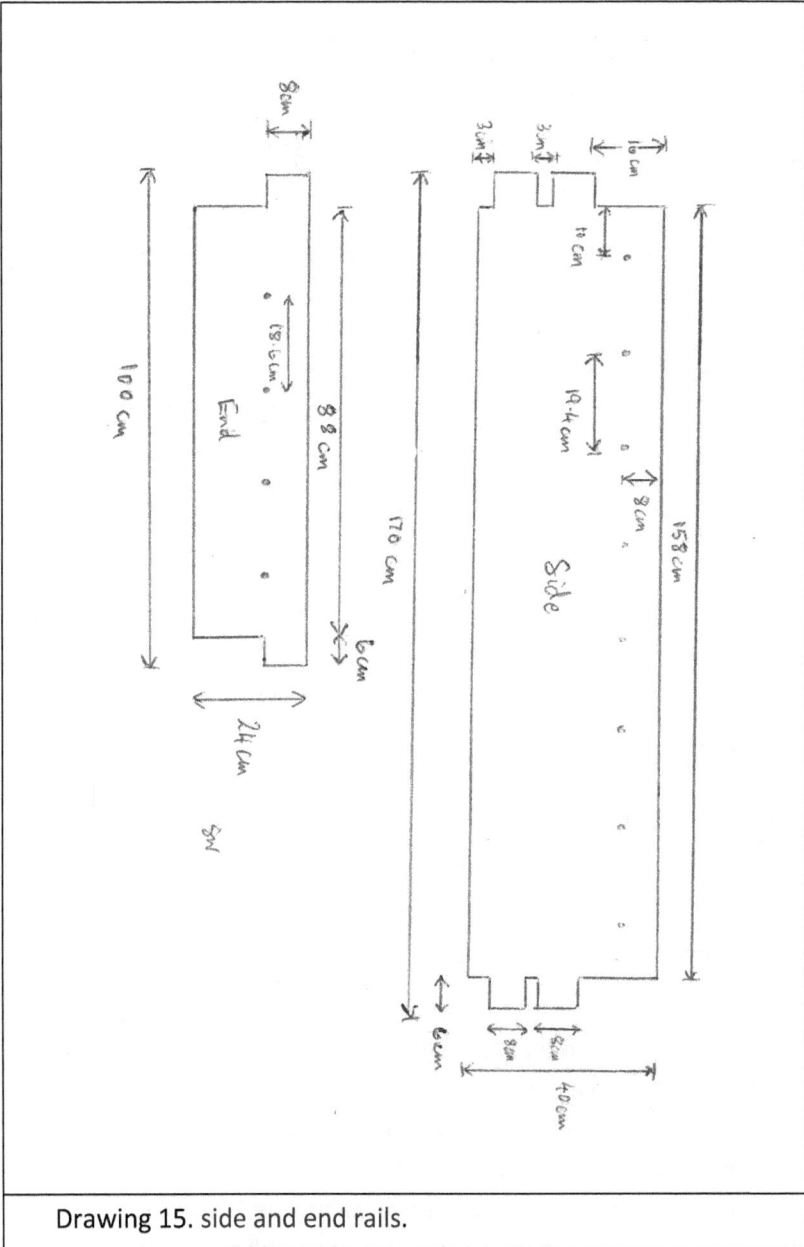

Drawing 15. side and end rails.

1. Mark out and cut sides and ends to the length of your desired mattress (allow for expansion and contraction of the timbers).
2. Mark out and cut out end tenons.

a) Drill 10mm (for 8mm rope, linen or hemp rope) holes in the sides and ends for the rope, as set out.

b) Round of edges of holes (so as not to damage the rope).

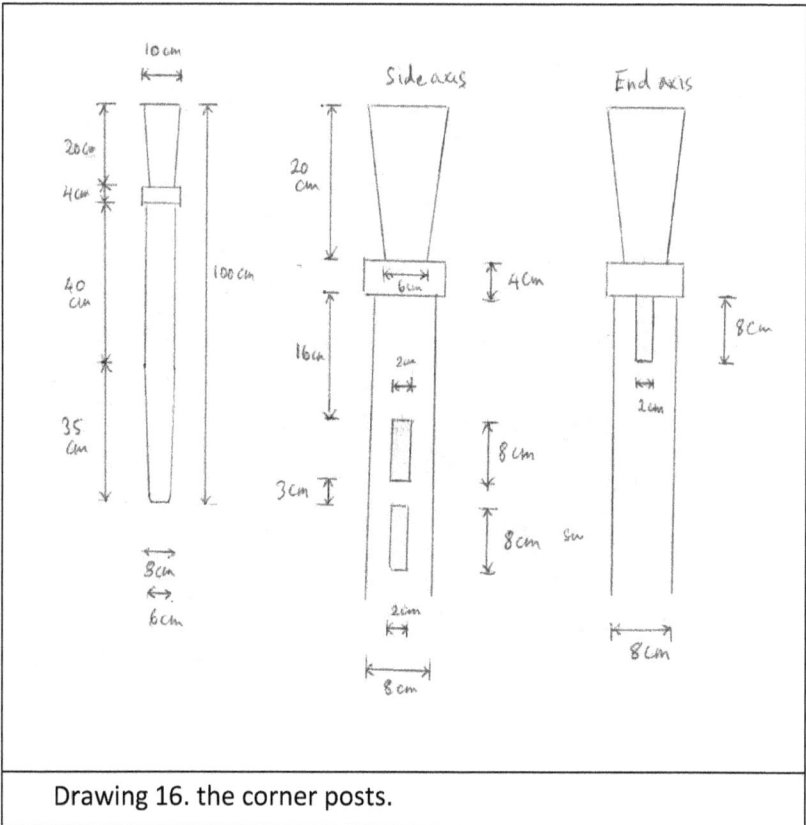

Drawing 16. the corner posts.

Legs.

3. Mark out and cut legs to length.

   a) Cut legs to length.

b) Mark out top and collar on all four sides.

c) Cut to depth on all four sides at:

   i. Just below and just above collar. See Photo 41.

d) Cut down from the top to top of collar on all four sides. See Photo 42.

| Photo 41. Horizontal saw cuts done. | Photo 42. Angled saw cuts down to the horizontal cuts. |

4. Shave down legs on all four side with draw knife. See Photos 43 & 44.

| Photo 43. Re-enactor from the New Varangian Guard Inc. – | Photo 44. Legs shaved down to size, tenons next. |

| Vlachernai branch shaving legs down to size. | |

Mortices.

5. Mark out the mortices on the legs. See Photos 45 & 46.
    a) Secure leg to work bench or vice.
    b) Drill out or chisel out the mortices.
    c) Clean up mortices with rasps and files.
    d) Ensure that there is a tight fit for the tenons.

| Photo 45. Leg on the bench secured by a 'hold down' ready to start chiseling. | Photo 46, Tenons in the legs are done. |

Fitting legs to sides and ends. See Photo 47 and 48.

6. Snugly fit side and end tenons into mortices of the legs. This may include taking of material from the sides, ends and the mortices of the legs. Check fit on a regular basis.

| Photo 47. Legs fitted to sides. | Photo 48. Legs fitted to ends. |

Identify the parts.

7. Carve ("*Marriage*") marks (i.e. Roman numerals) on legs and the ends of the sides and ends so it is easier to match up when assembling the bed.

Roping up the bed.

1. Tie a knot in one end of the rope and thread the other end through the first hole on a short side and across the bed and through the corresponding hole on the opposite side of the bed.
2. Then pass the end of the rope through the next hole on the side and repeat until you come out of the last hole on the side.
3. Pulling the slack through all the holes.
4. Run the rope around the leg to the first hole in the end, thread the rope through the hole and weave the rope over and under the ropes going from side to side.
5. Run rope through last hole and tie off. The rope may stretch with use of the bed so you may need to retighten the rope.

Note: You could leave the sides roped up for transport, and then the rope from the ends could be woven through the side ropes.

Mattress material can consist of a tick (a cloth bag) full of straw or some other natural material. Modern foam can be purchased and cut to size to fit snuggly inside in ends and sides.

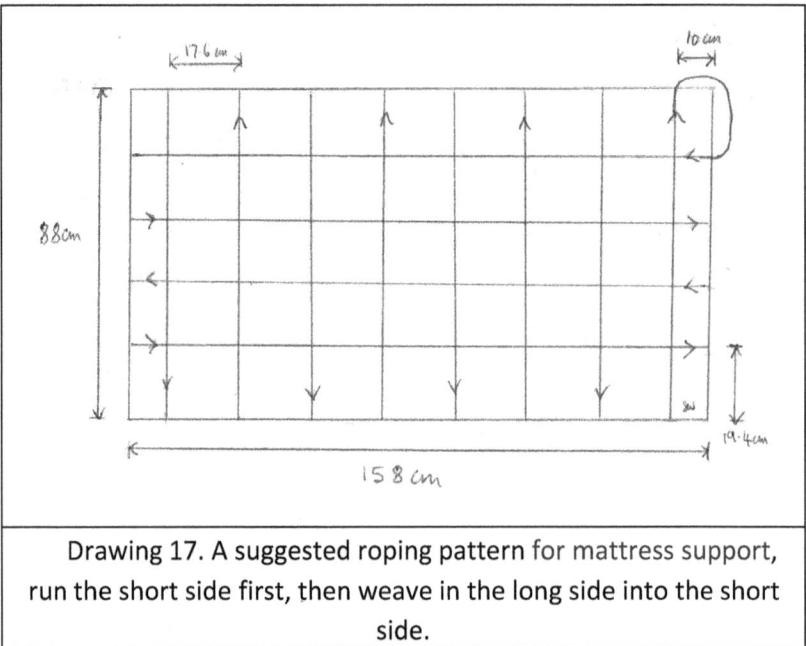

Drawing 17. A suggested roping pattern for mattress support, run the short side first, then weave in the long side into the short side.

Photo 49. The longer version of the bed in pine, with rope support for mattress.

The Table

The Sala Hytta table

The Sala Hyatt table	Grave 4, Sala Hytta, Västmanland, Sweden	When: 10 century
Stored: Statens historiska museer, Sweden.	Collection no: SHM 11739[20]	Material: Wood, not determined.
Size:	Table top: 45*20 * 2 cm	Height of table: 8.5cm

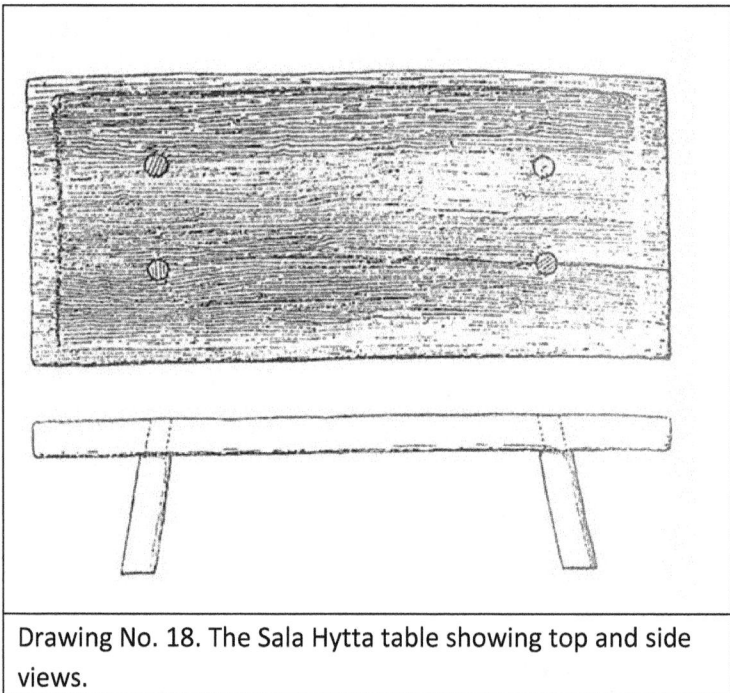

Drawing No. 18. The Sala Hytta table showing top and side views.

---

[20] Thanks to Gunnar Andersson (Senior Curator - Statens historiska museer – Sweden) for information regarding this item.

Photos No.50. Top of table, showing the chiselled out section the tops of the for legs.

Photos No. 51. The table from the corner of side of table.

Tables have been around for a long time, Egyptians, Greeks and Romans, all used them to keep things off the ground, for the purpose of storage, or eating or working at. Tables ranged in size, shape, materials used and types of construction, see table 13 for a comparison.

The 10th century grave 4 at Sala Hytta in Västmanland, Sweden, had a table which was 45 cm long, 20 cm wide and I have assumed the thickness to be similar to that of the Hørning table at 2 cm. The table stands on four 20.5 cm long legs which are 2.4 cm in diameter, the holes for the legs in the table top are only 1.5 cm in diameter.[21]

The construction techniques are very similar to the Lund stool except for the increase in the number of legs and the slight indentation of the upper surface. This indentation could be for insuring that objects on the table did not fall off easily, or to retain a certain amount of spilt water if used as a wash table.

In Drawing 20, it can clearly be seen that the legs are splayed towards the long axis, I suggest that the legs should also splay towards the short axis (which would increase stability).

Table 13. Table comparison.				
Place	Date	Collection	Material	Dimensions
Castleford, Yorkshire, England	1st century CE	To be determined	Unknown	L. 61cm, W. 30cm, thickness 2.7cm, only top found.
Oseberg farm, Tonsberg, Vestfold, Norway	800 – 850 CE	Universitetets Oldsaksamling, Oslo, Norway.	Oak	L. 92.5cm, W. 33cm, Legs from 28 to 29cm
Hørning Church, 40 k east of Mammen, Denmark	950 CE	To be determined.	Unknown	Top 50cm square with edging, on 20cm high legs.

---

[21] There was a similar table found in the Roman fort at Castleford, made of ash, measuring 60*30cm with four splayed holes for legs, the legs did not survive.

Materials.

The table top, timber plank, 45*20 * 2 cm,

4 Legs, timber dowel, 20.5 cm * 2.4 cm,

Wood glue, Aquadhere is recommended.

Tools.

For a more authentic table I suggest you avoid any power tools and use wood working tools of the period, see the tools of the Mästermyr tool box (Arwidsson and Berg 1983). Or a protractor, pencil and ruler, hand saw, hole cutter or spoon drill, chisel and mallet, rasp or knife (or a draw knife or spoke shave).

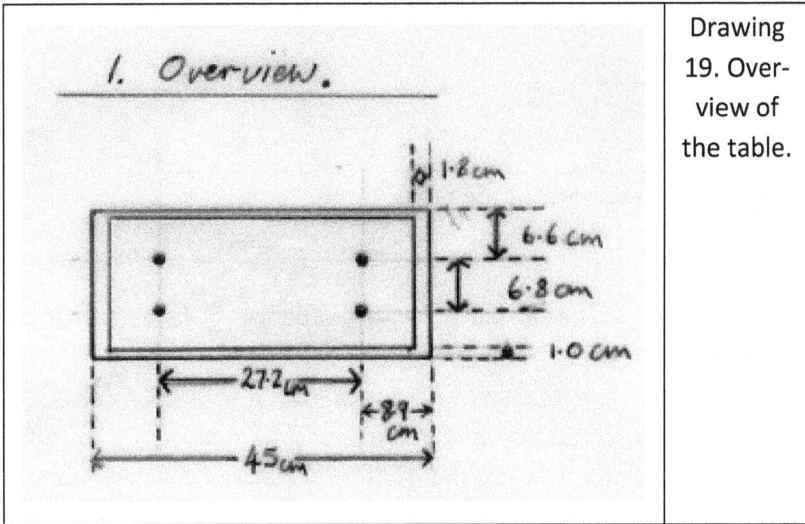

Drawing 19. Overview of the table.

Construction Instructions.

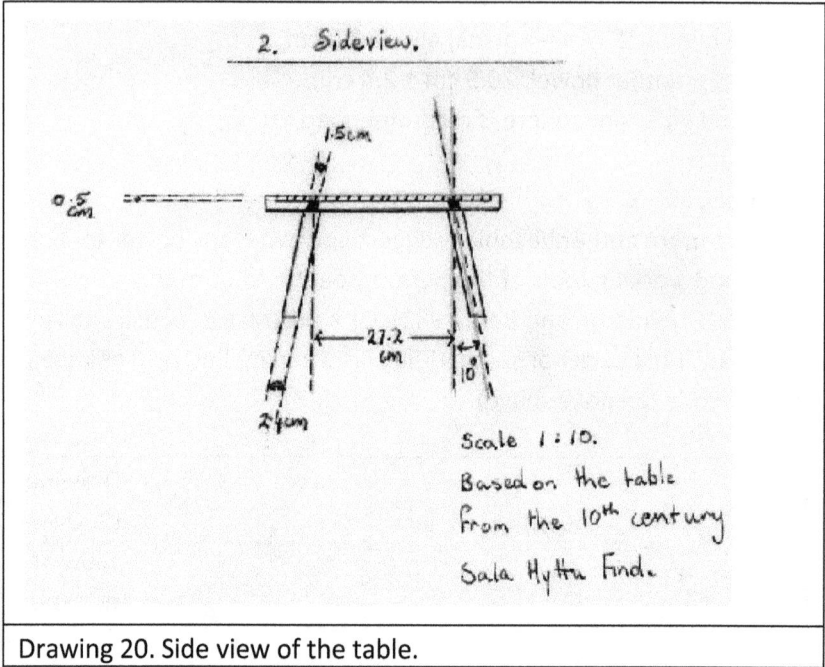

Drawing 20. Side view of the table.

1. Select a suitable piece of timber for the table top.

2. Mark out timber as per plan using a pencil. Make sure to avoid cutting through knots.

3. Cut timber to size using a handsaw.

4. Cut leg holes 15 mm wide in the appropriate places on the table top. Angle holes at 10° from the horizontal plane of the table top towards the closest corner of the table. See Drawing 20, Photos 52 & 53.

5. Gouge out sunken area to an even depth of 0.5 cm.

6. Cut four lengths of 2.4 cm dowel in 20.5 cm lengths. Cut ends at a 10° angle to match the table top and the ground.

7. Rebate the tops of the legs to fit into the holes in the table top. The legs must be a snug fit or they will make the table unstable. Remove excess leg if poking up through the hole. A sloping shoulder could also be made on the legs if you wanted to remove the legs for flat packing. See Photos 54 to 58.

8. When legs are a snug fit smear wood glue around top of leg and slide into place.

9. Clean off the excess glue and allow to dry.

10. Plane, scrape and clean up table.

11. Place name and group on underside of table top. This measure helps distinguishing your table from anyone else's.

Photo 52. Drilling holes for legs with drill, spade bit and jig.

Photo 53. Chiselling out the insert in the table top.

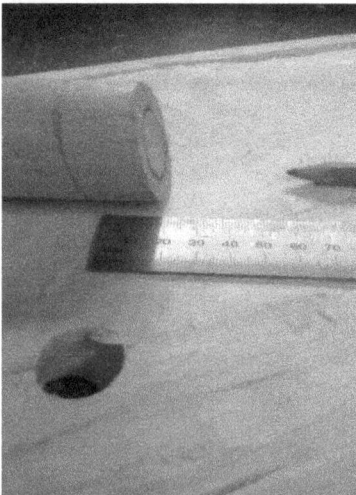

Photo 54. Marking the leg.

Photo 55. Cutting the tenon.

Photo 56. Chiselling the tenon.

Photo 57. Rounding off the tenon with a rasp.

Photo 58. Fitting the legs.

The Toy

The Trondheim toy horse.

The Trondheim toy horse	Trondheim, Norway	When: c.1075 - 1125
Stored: Vitenskaps-museet Trondheim, Norway.	Collection no: N97259/FU450	Material: Wood, undetermined.
Size: 120mm * 51mm * 11mm. [22]		

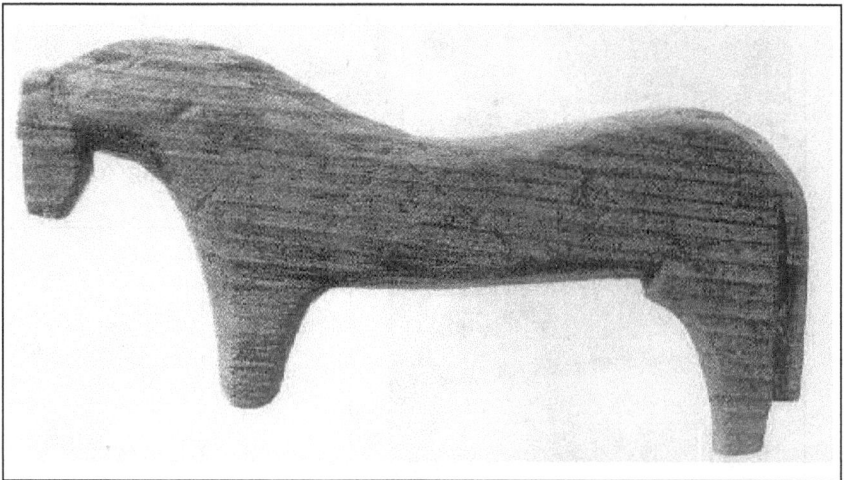

Photo 59. The original Trondheim Viking toy horse.[23.]

---

[22] Thanks to Jenny Kalseth (Norwegian University of Science and Technology) for the information on dimensions of the horse.

[23] P. 231 Roesdahl, Else, and David M. Wilson. From Viking to Crusader: the Scandinavians and Europe, 800-1200. New York: Rizzoli, 1992. ISBN: 0847816257, has the length of the horse at 12.7cm.

Table 14. Other Viking toy horses.	
The St. Petersburg (Staraja Ladoga)	*Wood, L. 13.0 cm, Staraja Ladoga (horizon D) obl. St. Petersburg, Russia, 8th - 9th c., Gosudarstremyj Ermitaz, St. Petersburg, LD-244.*
The Faroes Viking toy horse	*Fir, L. 13.2 cm, Kvivik, Streymoy, Faroes, Viking Age, Local, Føroya Fornminnissavn, Tórshavn, 4765/3797.*

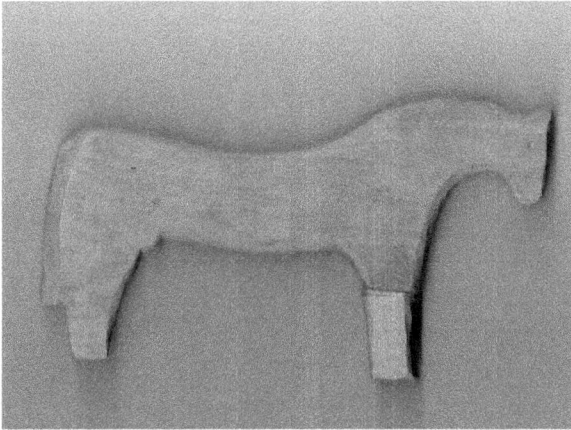

Photo 60. Replica of Trondheim Viking toy horse, with broken leg (made from Huon Pine).

Children of the Viking period had a number of different toys to play with; wooden swords, dolls, boats, ducks, cats and horses. The Trondheim toy horse dates from c. 1075 – 1125 and represents household production for use in the home as do the other extant toy horses (i.e. St. Petersburg and the Faroes), see Table 14. It should be noted that all three are close to thirteen centimetres in length, this could mean that such carving has a specific origin and the size is a continuing tradition even though the design has been modified.

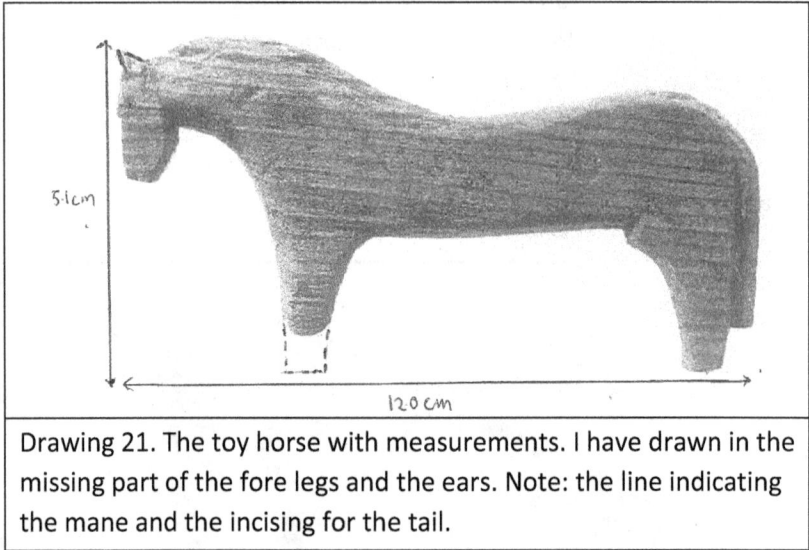

5.1cm

120 cm

Drawing 21. The toy horse with measurements. I have drawn in the missing part of the fore legs and the ears. Note: the line indicating the mane and the incising for the tail.

In reference to the damage on the original Trondheim toy horse that my son Bryn has carried out some tests on his replica and has produced similar damage. Bryn's use of the horse as a teething ring and the odd bash on the polished floor boards of our house has resulted in numerous teeth marks especially on the legs and head, and a broken foreleg.

Materials.
Timber, preferably a soft wood, a pine or a fir.
Out of the three toy horsed discussed here only the wood for the Faroes horse has been identified as willow. So I suggest you use willow, fir or any other similar soft wood found in the region and in the period. I actually used Huon Pine (Tasmania) since I had a plank handy.
Do not coat the toy horse with any toxic substance because this may end up in someone's mouth, such as a child's.

Tools.

Axe, a sharp back bladed knife, whetstone to sharpen knife.

Construction Instructions.

Note: It has been assumed that the carver has some notion on how to carve wood with a knife.

Precautions. When cutting towards yourself don't slip or alternatively secure the work with a clap or vice. Wear a leather apron.

Version 1. (The old fashioned way).
1. Split log into planks, dress to about 1.5 to 1.0 cm thick.2. Mark out outline of toy horse on one side of a plank. 3. Carve horse out of a plank. 4. Carve instep on each side of the tail. 5. Carve horse's mane. 6. Use the point of a knife in a twisting motion to add depressions for the eyes.

Version 2. (A faster version).
1. Dress a plank of wood to 11mm thick.

2. Mark out outline of toy horse on one side of plank with the grain running parallel along the body.

3. Cut from plank using a coping saw or an electric jigsaw (use appropriate safety gear). Or slowly whittle to shape using a sharp knife, cut away from your body and hands.

4. Carve shape the main outline of the horse, Avoid applying pressure laterally on the legs because they will snap.

5. Cut a line along the top of the neck for the mane, see line in Photo 60.

6. Cut two parallel lines down from rump to about a centimetre from the bottom of the rear legs, about 8mm apart. See Photo 61.

7. Cut a horizontal line across the tail across the rear legs about a centimetre from the bottom of the rear legs, about 8mm apart, cut off excess. See Photo 61.

8 Saw a cut down through the ears (a couple of mm), then file a groove between the ears, See Photo 62. Note: the ears are missing from the original.

9. Cut details around horse's genitals.

10. Finish off using the tip of a sharp knife to twist in eyes on either side of the head.

11. Ensure the bottoms of the legs are flat so the horse is free standing. This can be done by rubbing the horse's hooves over a piece of sand paper on a hard surface.

Photo 61. The Tail.

Photo 62. The ears and eyes. There is not much of the ears left on the original but this is how I made mine. I also used the point of the knife in twisting motion to create eyes, so it could see.

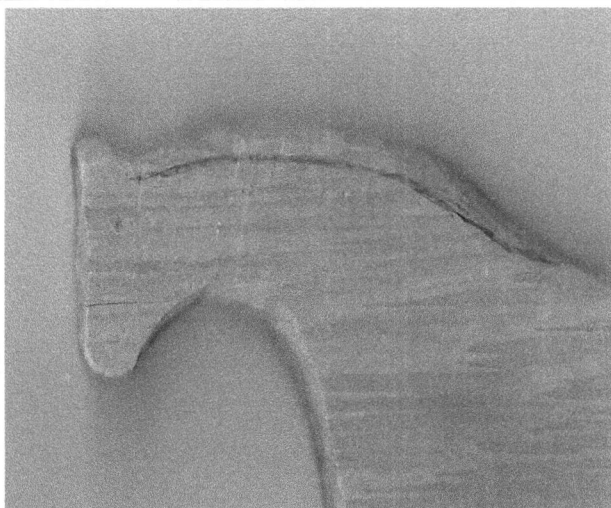

Photo 63. The mane. There is a line on the original which may indicate the depiction of the horse's mane. I have drawn in pencil approximately where the line travels from the ears to the shoulder.

Candle Stands

The Gokstad candle stands by Wayne Robinson (Ret[d]).

The Gokstad candle stands	Gokstad burial mound	When: 900-905CE
Stored: Kulturhistorisk museum / University of Oslo	Collection no: un-known	Material: oak
Size: various, 15-17cm+		

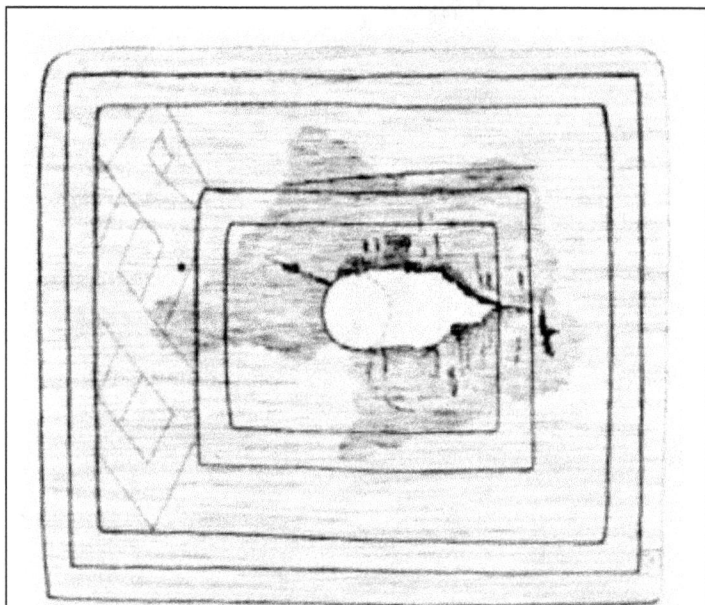

Photo 64. Figure 5 from Nicolaysen, the large candle stand.

Photo 65. Picture of replica by Wayne Robinson.

This project is a good one for a beginner, there are a few options so you can make it as simple or as complex as you want to.

Four small wooden boards were found in the Gokstad burial mound (c. 900-905CE), in the aft of the ship along with a cauldron and other kitchen equipment. Each of the boards has a round hole in the centre to hold the base of a candle. Recent analysis has identified a local origin to the find site and dated them to a few years prior to the construction of the mound.

The original dig report is a little light on details, being from a period where the finds are regarded as antiquarian curios. We know the shapes and some dimensions: two 150mm diameter circles; one 170mm square with the corners cut in reverse semi-circles and; the rectangular one with the carving. The thinnest is 5mm thick, the others somewhat thicker. We don't have exact dimensions for the carved stand. The report has a scaled drawing with a comment that it is ⅓ scale, but I've never seen an original copy of the book, and all the digital copies floating around in the wild are not the original printed size so we can't just measure off the screen and multiply by 3. The ratio of short side to long side is 6:7, so if we assume the carved stand is 150mm wide to put it in the same approximate size range as the others originals, it will be 175mm long and have a 20mm diameter hole for the candle.

We'll concentrate on the largest one here, as it is the most complex, but all techniques used could be easily applied to the simpler ones. You'll note that my stands in the photos below are somewhat squarer than they probably should be. That's because they are made from waste timber – the cheeks that I'd sawn off a tenon for a bench I was making. Remember to drill the hole the correct size for the candles you'll be using.

Materials.

Wood – oak is perfect, but other hardwoods will work. Due to the structure of the cell walls, hardwoods are considerably easier to carve than softwoods.

Candle – a candle of the size you wish to use and can get more of.

Drying oil of choice – linseed oil is easiest to get, but walnut or tung oil will work too. Boiled linseed is safe for this application. Don't use cooking oil, it will never dry.

Thinner for the oil – gum turpentine if you're using paint, or plan to eat from the candle holder, otherwise mineral turps is fine. Both are in the same section of the Big Green Hardware Store for considerably less than you'd pay at the art supply store, just without the tastefully monochrome picture of the naked lady on the front.

Paint or pigments if you're using them – I'm assuming oil-based paints, you may need to change the order of oiling and painting steps to suit the type of paint you're using if it's in a water-base. Check the label for application instructions.

Tools.

Minimum: Handsaw, drill with candle sized bit, clamps, pencil, sharp knife, metal ruler, scrap timber for backing the drill, PPE.

Optional: hand plane, shooting board, marking gauge, marking knife, cabinet scraper, v-gouge, mallet, abrasives, paint brushes, shiny glazed tile, polished pebble that isn't uncomfortable to hold.

Step 1 – Assuming you didn't miraculously find a piece of timber the right size, mark the desired size and cut it to length. Cut it to width, too if you need to. Clean up the ends with sanding, files, or a plane. I'll wait while you do that.

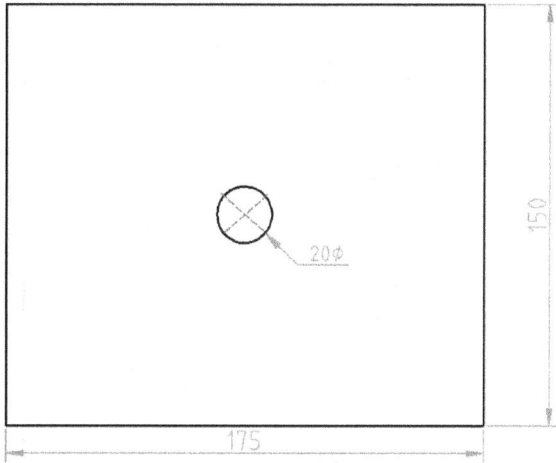

150

20⌀

175

Step 2 – Mark the centre by drawing short lines from the diagonals. The shorter and closer to the middle you make these lines, the less there is to clean up. If you keep them completely within the diameter of the candle hole, there's no clean up at all.

Clamp the stand to the workbench[24], with another piece of hardwood behind where the hole you are drilling will emerge. This will avoid chipping the back of the hole when the drill breaks through. The wood is too thin to be able to drill from both sides cleanly, which is how I usually avoid tear-outs on the surface. Make sure the backing piece is thick enough that you won't drill into the table, especially if it's mum's good dining table.

---

[24] For your given value of "workbench". It could be anything from your dining table to a manky bit of ply from the neighbours' hard rubbish balanced on a milk crate.

Drill the hole in the centre and clean up the edges of the hole as required. If you make an absolute bollocks job at this point, either start over, or char the edges of the hole with a candle so it looks like it was meant to be that way all along. Don't ask me how I know.

Step 3 – Carve the frames. Mark points an arbitrary distance in from the four corners that looks right to you, then draw a line between them. I used ¼ of an arbitrary king's thumb[25], but about a barleycorn[26] is probably right.

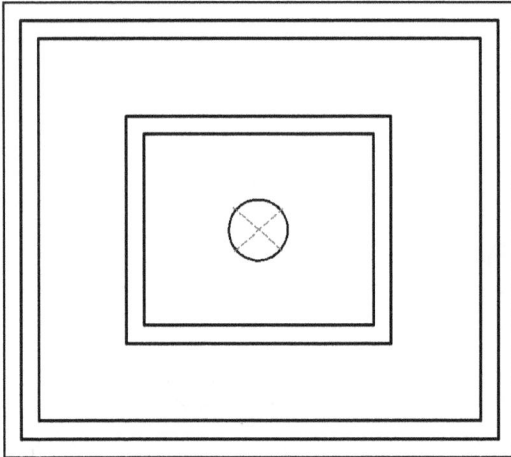

If you have a v-gouge you already know how to sharpen and use it. If not, the quickest and neatest way to do the carving is with a Stanley/Ir-win/Craft knife and the metal rule. Do one vertical cut along the line with firm pressure, then move a little to either side and do an angled cut aiming at the bottom of the vertical cut. You may need to run the knife along the cuts a couple of times to get the waste to come out cleanly. Don't pick at it, because that will leave chips and ragged edges.

For the love of all that is holy, KEEP YOUR FINGERS BEHIND THE EDGE OF THE RULER. Your thumbs may grow back, but oak reacts with the iron in blood and stains black. Guards are available at craft shops if you don't think you can cut safely without one.

I strongly suspect this was originally the only decoration, and the knot work was done by a bored Viking, either with a view to

---

[25] 0.25" = 6.35mm

[26] An archaic measure equivalent to ⅓", 8.46mm or 4.20875x10⁻⁵ furlongs, for those of you using the FFF system.

improvement or overcome with *horror vacui* and a surfeit of mead. If the border was the only decoration, that would make it similar to the Hedeby sea chest which also has a framework of pairs of parallel lines.

Step 4 – Mark up the knot work, if I haven't dissuaded you by this point. You need to decide whether you'll copy the knot work from the original candle holder or whether you'll do it properly. Now decide if you'll copy the mistakes on the original, or if you're willing to make your own.

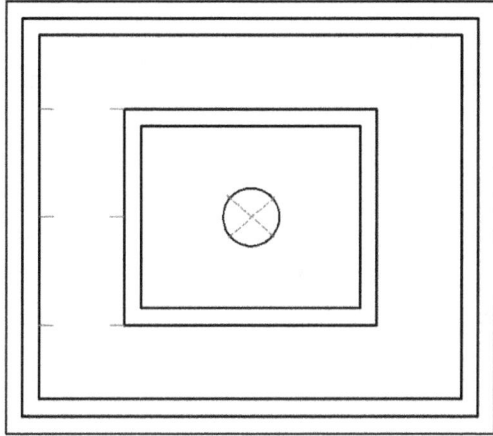

Start by dividing the short side in four on both sides of the frame.

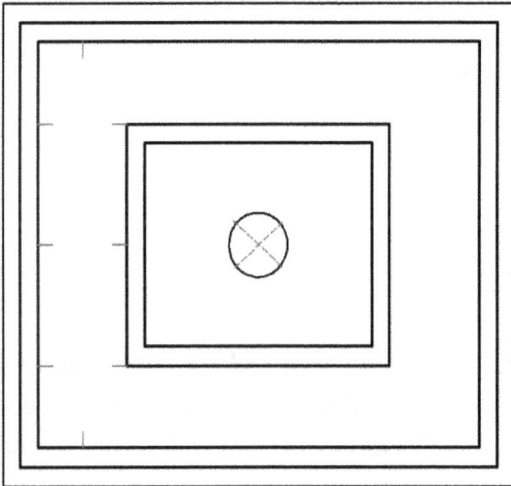

Now mark the midpoints and extend them to the borders. These are where the knotwork turns back on itself, and should be in line with the middle of the wide space. Yeah, that's confusing. Look at the picture, those are them a little way in from the corners on the two long sides.

As the knotwork doesn't meet the frames at single points on the sides, you'll need to add some width. Measure a distance either side of the side midpoints, I used ⅛"/3mm either side to get a total width of ¼"/6mm for the knot-work bands. If you're us-ing different width knotwork, you'll need to add half the width of your knotwork either side of the midpoint. The ends of the knotwork do meet the carved border at the points, so you don't need to worry about width for them.

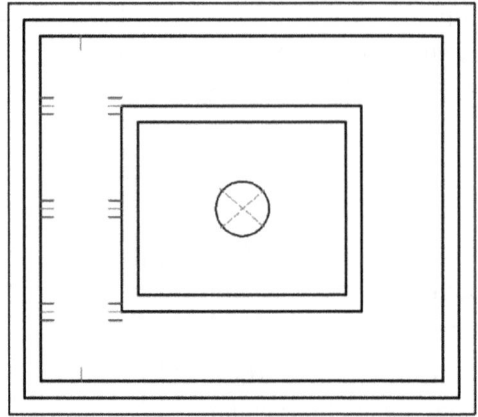

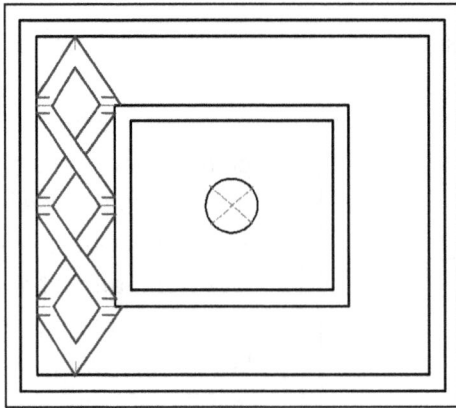

Join the dots, paying attention to which band goes over which. The inner edges always come to a point in line with your original marks, the outer lines go to the wider marks so there's a flat where the knotwork hits the carved frame on the short sides.

Step 5 – Carve the knot work. Yeah, I cheated and used a marking knife to do the mark up. Mine's mostly done and just needs a little clean up with the V-gouge. Don't worry if it isn't perfect, errors add character, and remember Benni Forkbeard couldn't keep his inside the lines.

Photo 66. Picture of replica by Wayne Robinson.

Step 6 – If you are using water-based paints, either skip this step, or do it after step 7. The paint instructions may direct you to use a sour milk wash first, and then wax after the paint is dry.

Oil the whole lot. I usually mix some inna jar and use a brush, but some people prefer to wipe it on with a rag. Research the correct way to handle and dispose of oily rags if you're going to do it that way[27]. The brush will wash out in turps.

A 50/50 mix of linseed or boiled linseed with turpentine will seal and protect the timber. You need to use gum turpentine rather than

---

[27] Spontaneous combustion is a thing, I've seen it a couple of times.

mineral if you're planning to paint over the carving, or the paint will crack and flake. Put it outside for a good dose of UV to help it set and leave it until dry.

Optional step 7 – You can paint if you want to. We can leave your friends behind. There's no evidence of paint on the originals[28] so we're in the wild world of supposition here.

If you're grinding your own pigment, get about ¼ tsp of pigment and add two drops of oil. To get each particle of pigment coated with oil, you need to use shear force. Gently rubbing the pigment and oil on a glossy tile with a polished pebble should be enough. Keep adding tiny amounts of oil and/or gum turpentine and keep rubbing until its flowing right.

Carve some grooves in a scrap bit of timber and keep testing on them until it flows and covers the way you want it to. You'll get a feel for it fairly quickly.

If you're using oil paint from tubes, squeeze a little out on a saucer and add a drop or two of the gum turpentine to the right consistency. A china saucer from the op shop works well as a palette, don't use Nana's Good Crockery™ for this. The test piece of hardwood works here for getting the flow just right, too.

If you're using milk, chalk, tempera or poster paints, you probably already know what you're doing and don't need me to tell you how to separate the membrane from the yolk.

Paint the grooves, paint the raised fields to taste This is your little bit of wood and you can paint it how you want.

Colours – some of the other wooden items on the ship were painted, and the pigments have been analysed. There is yellow from orpiment, red from iron oxide, white from chalk, and carbon black. I used yellow ochre instead of orpiment, life's too short to grind arsenic

---

[28] I'm not entirely convinced that traces of paint would have survived, or that if it had, anybody looked for it on the smaller items. They were altogether too smug about having found the boat, shields and weapons.

(and likely to get shorter if you muck around with heavy-metal pigments too much). Have a look around at other objects like painted stones, churches, arrows, and other burials to see what other colours you can use.

Disclaimer: Australian Consumer Law requires that candle holders sold in Australia pass the "combustibility test" requirements of the ACCC's CPN #12.

Bag Hanger

Birka bag hanger
by Shannon Joyce

Bag Hangers	Find location: Lake Malaren off the island of Birka.	Dating: 700 to 900 CE.
Held: Maritime History Museum, Stockholm	Collection no: F 153	Material: Wood, species not known
Size: 282 mm* 50 mm * 7 mm		

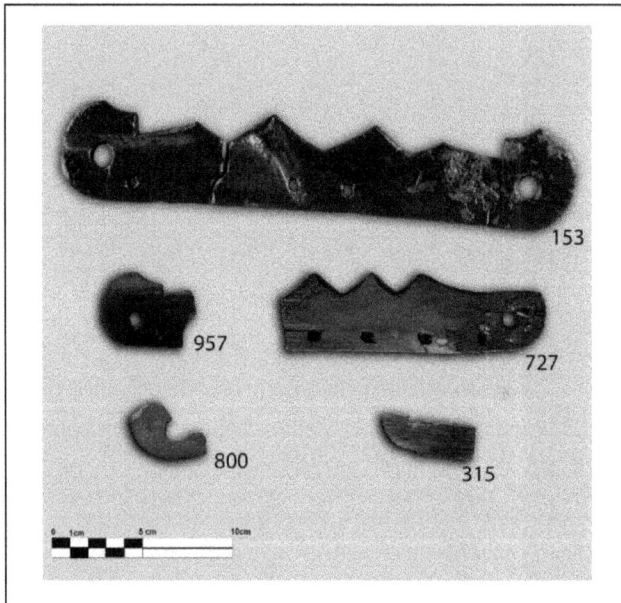

Photo 67. Bag hanger parts found in the waters off Birka. Photo by Christin Mason, Swedish National Maritime and Transport Museums, Stockholm, Sweden.

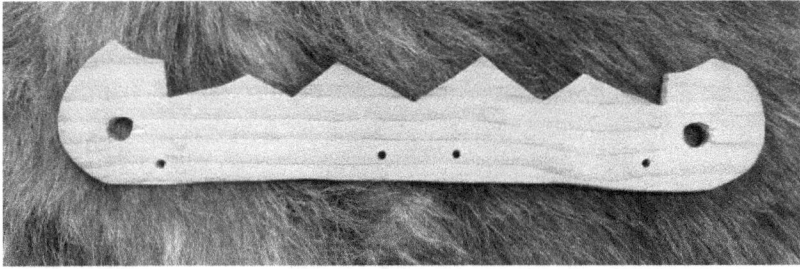

Photo 68. Completed hanger based on Birka F153.

This style of bag is ubiquitous among Viking age reenactors world-wide. Go to any festival or event and you will see both sexes carting their feasting gear and mundane necessities around in these bags. Called the Hedeby Bag after the site of most of the finds, the form that this bag takes is a combination of replication of extant hangers and conjecture regarding the shape and material of the bag hanging from them.

In this chapter I will look at the extant hangers and gather clues from these and other sources to speculate what form the bag took. I will show step-by-step how to make a set of replica Birka F153 hangers and a simple wool and linen bag to suspend from these. I will also discuss a possible alternative structure using clues from contemporary art and experience using the bag as it is conventionally reproduced over a period of two re-enactment seasons.

Hangers from Birka.

A total of seven items were found underwater in Lake Malaren off the island of Birka in 2014, just outside of Stockholm, Sweden, see Table 15 for details. Three of the fragments are part of the same complete hanger; F 153 (Photo 67), two are parts of the same incomplete hanger whilst the remaining finds are parts of frames.

Table 15. Details of bag hanger finds from Birka. Source: Swedish Maritime Museum, Stockholm.[29]				
Find No.	Item description	Length mm	Width mm	Thick-ness mm
153	Complete hanger in three parts.	282	50	7
315	Possible fragment of bag hanger	53	18	7
727	One end of bag hanger	145	44	6
800	Part of bag hanger	45	30	3
956	Fragment of bag hanger	56	40	7

Structure of the Bag.

There are no surviving complete hanger bags from the Viking age so any construction that we use is going to be hypothetical. The standard construction that is seen is a square around 30cm suspended from the hangers with a long cord threaded through the hangers to form a shoulder strap. This construction is perfectly reasonable and is one I have made and used but have found to be problematic as I will discuss later in this article.

So what can we look at to get a sense of how the bag could have been constructed?

---

[29] I am unable to find information at the date of publishing on what species of wood they are made from.

Looking for Evidence.

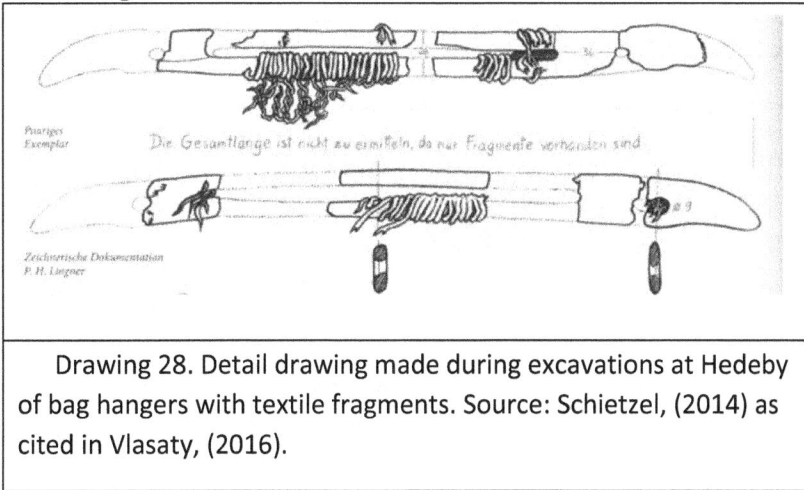

Drawing 28. Detail drawing made during excavations at Hedeby of bag hangers with textile fragments. Source: Schietzel, (2014) as cited in Vlasaty, (2016).

Textile remains on hangers.

Hangers from Sigtuna and Hedeby show that many threads were used to attached hangers to bag (Drawing 28). There is also a clue as the construction of one of the bags; the textile fragments show twisted threads, sprang[30] perhaps, attached to the hanger. Unfortunately the textiles depicted in these drawings have not survived to the present day (Vlasaty, 2016).

Iconography.

There are many examples of bags from the middle ages to be found in iconography (see Table 16), however, we need to be mindful of the fact that not a great deal of this comes from prior to the 11th century. Those examples we do have depict similarities include the way of wearing of the bag; across the body on a long cord and that, as far as can be concluded taking artistic representation into account, they don't appear to be overly large or deep. These are practical, personal

---

[30] Sprang is a type of textile, created when warp threads are systematically twisted around one another, and as a result looks a bit like netting, except it's not looped or knotted.

accessories designed to carry items that are too big or numerous to fit into a pouch.

Table 16. Examples of frame bags in iconography.	
	Figure 2. Tacuninum sanitatis, 47v., detail. Late 14th century. Frame bag with gathered sides allowing for large top opening. Shallow depth. Shoulder cord is a loop threaded through the ends of the frame and gathers on side of bag. Bag frame appears to be internal.
	Figure 3. Tacuninum sanitatis, 29v., detail. Late 14th century Another example of gathered sides with internal bag hangers. Bottom corners and centre of the bag fabric look to be bound together in a knot.
	Figure 4. Detail of The disrobing of Christ from the Karlsruhe Passion by Hans Hirtz c.1440. Long cord is looped through the internal hangers and the gathered sides of the bag.

	Figure 5. Detail of one of the frescos to be found in the crypt of the Saint-Nicolas Church in Tavant, France. Dating from the 12th century. A cross body satchel on a pilgrim.
	Figure 6. Harley Psalter 66v. First half of the 11th century Cross body bag on a man receiving alms from the King.

Other Cultures.

The use of carved bone and antler bag hangers by the Sami of Northern Scandinavia and Russia also provide clues as to the possible construction of the bags and how the hangers were used.

	Photo 69. Sami bag made of leather and reindeer horn hangers, edged with red woollen fabric. Wider than the hangers and shallow to allow for access to contents Image Source: Norske Folkmuseum https://digitalt-museum.no/011023294759/veske?page=71&pos=1685&query=veske[31]

---

[31] Free use via a creative commons licence deed by SA 4.0.

We can speculate from the evidence several things about the material and construction of the bag and how it was attached to the hanger:

1. Hangers could be internal;
2. Shoulder cord could be threaded through hangers and sides of bag;
3. Could be shallow;
4. Plain hangers with smooth edges facing upwards.

Making Viking Bag Hangers.

To make a pair of Birka F153 Viking Bag Hangers you will need:

Materials and tools.

Timber (I used a piece of pine 90 x 65 x 9mm), linseed oil (boiled) and mineral turpentine, paper and pencil. Ruler, coping saw or an electrical jigsaw, drill with 10mm and 5mm timber drill bits, camps, sandpaper and files.

Construction Instructions.

The dimensions of the extant hanger are 282 x 50 x 7mm. Draft a pattern or enlarge a copy of the extant hanger using a photocopier until it is approximately the right size. Cut out and use as a template to trace the shape of the hanger twice onto your piece of timber.

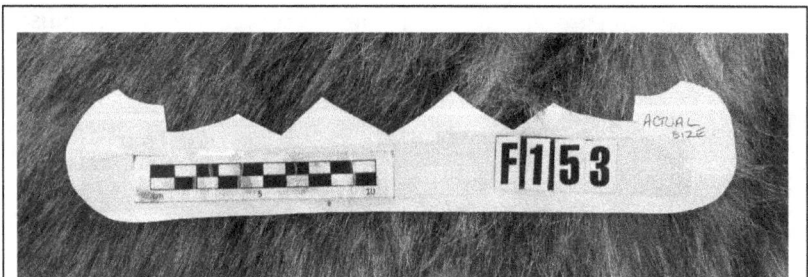

Photo 70. Template of Birka bag hanger F153.

1. Slowly and carefully cut out the shapes using an coping saw or an electric jigsaw.
2. Once the hangers are cut out clamp them together and mark where the two larger and four small holes are placed
3. Drill the larger holes with the 10mm drill bit and the smaller ones with a 5 mm drill bit. Unclamp.
4. File rough edges and give both hangers a thorough sanding.
5. Treat the hangers with an equal parts mix of linseed oil and mineral turpentine. Allow to dry for a day before repeating.

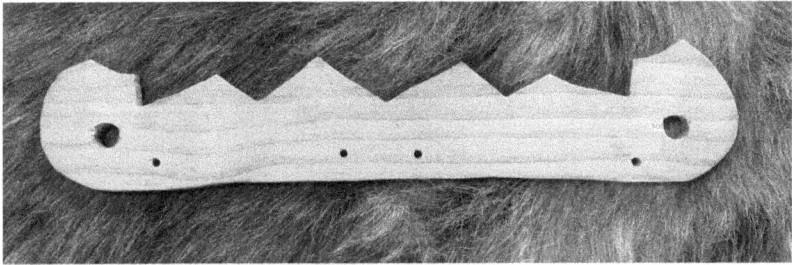
Photo 71. Completed hanger.

Making a Bag – Conventional Design.

This is an easy and plausible design for use with your bag hangers.

Materials and tools.

Outer fabric piece measuring 35 x 70cm (A heavy wool is suitable (in a period colour and weave), lining fabric piece measuring 35 x 70cm (linen), linen thread and sewing needle, 1mm cord or twine for attaching hangers (wool, linen, jute or hemp), 1.5m of 8mm cord or rope (wool, linen, jute or hemp).

Construction instructions.

1.

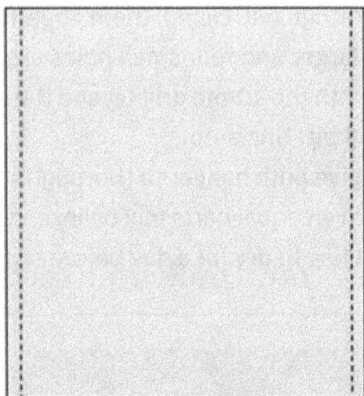

Fold the rectangle in half with the right sides of the fabric together. Sew the sides together to form a pocket.

2.

Turn the lining pocket right side out and press seams.

Construction instructions.

3.

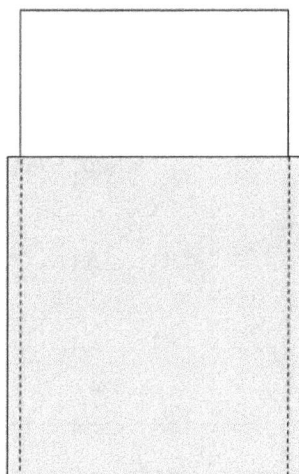

Insert the lining into the outer pocket with right sides of fabric together. Pin.

4.

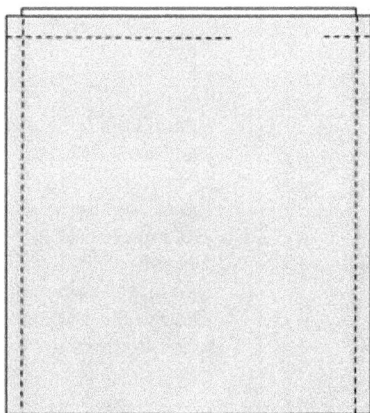

Sew around the top leaving 10-15cm for turning.

5.

Turn your joined pockets right side out and push lining into the outer pocket. Edge stitch the opening closed. Press.

6.

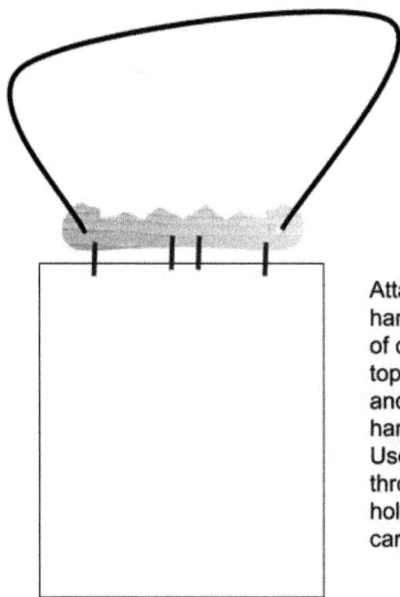

Attach your hangers with loops of cord through the top of your fabric and small holes in hanger.
Use larger cord through the end holes to make a carry strap.

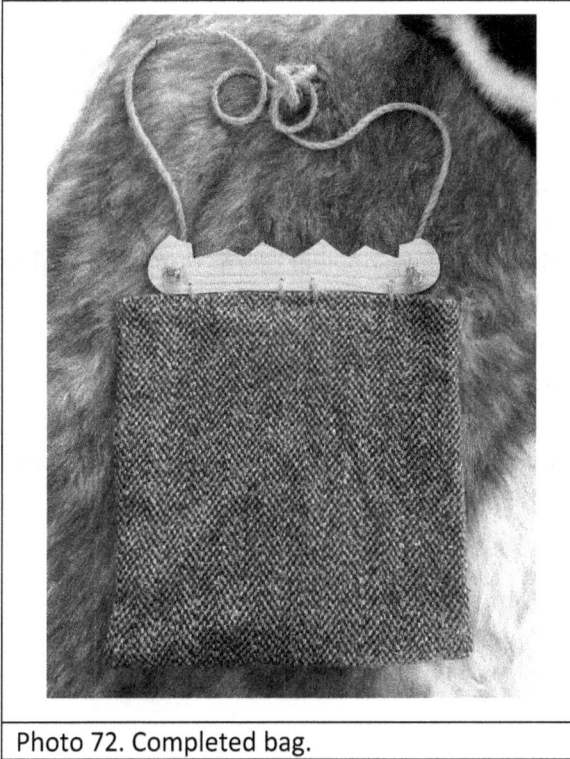

Photo 72. Completed bag.

Notes on an alternative construction.

This was my first construction of a Viking Age hanger bag and using it I found it to be problematic; the bag seemed to dangle loosely from the hangers and after a while the cords that held it came loose and needed to be repaired. This instability combined with a narrow top opening made it difficult to access contents; I would have to steady the bag with one hand whilst rummaging around in it for feasting gear, fire lighting equipment etc. The opening at the top of the bag never seemed to be big enough to be able to see what I was reaching for even though I had made the bag wider than the hangers on each side to allow for this. I have seen other constructions use side gussets to allow for easy access. I also scratched my hand and bent fingernails on more than one occasion when reaching inside.

My next attempt at making a hanger bag I will endeavour to overcome these issues by:

- Using a longer, plainer style of hanger.
- Patterning for loosely pleated sides to allow a greater opening as in Plates 1, 2 and 3.
- Making the depth shallower as in Plates 1, 2 and 6.
- Running the shoulder cord through the hangers and pleated sides (Plates 1, 2 and 3).
- Making the shoulder cord longer to allow for wearing across the body (Plates 5 and 6).

My aim is to produce a plausible, practical design for an item based on the available evidence; the aim of living history re-enactors everywhere.

In conclusion.

Of all the items mentioned in this volume I have only seen the Oseberg remains up close. I look forward sometime in the future to actually see the other items up close and personal to see the chisel marks, etc., so I can make any refinements to the plans and instructions in this and other volumes.

This is the first of a number of planned volumes on constructing historical items, the next off the rank is a volume on 14th century items but other volumes have been drafted on Byzantine, Crusader and the 16th century, and yes, there have been other volumes planned for Viking equipment.

Good luck, have fun and be safe.

Appendix 1 - Oseberg 178 chest

Larger version.

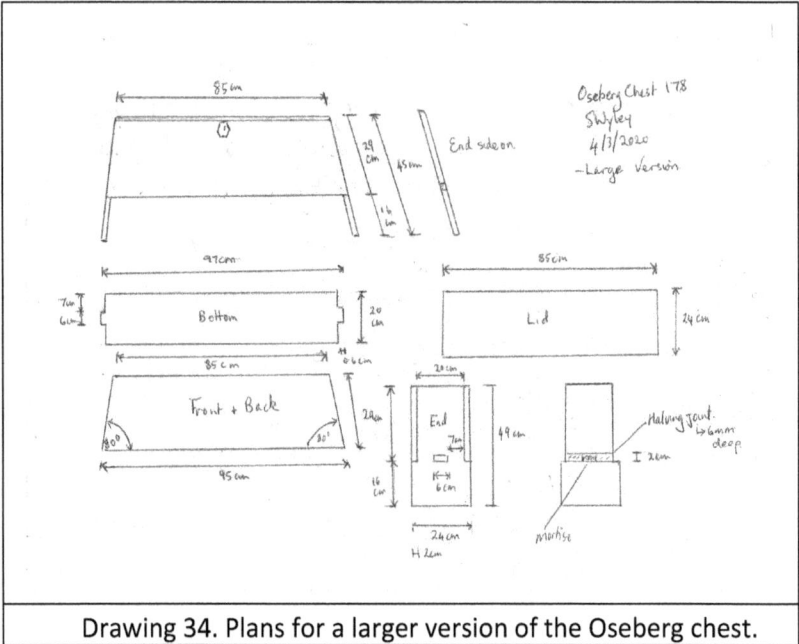

Drawing 34. Plans for a larger version of the Oseberg chest.

Appendix 2 – Oseberg 178 chest

Hasp and hasp plate details.

Table 17. Metal requirements			
Steel	Hasp and hook	Hinges	Hasp plate
Mild Flat bar (3mm)	230mm	1330mm	X
Mild sheet (2mm)	X	X	80*70mm
Wire (2mm)	X	320mm	X

Hasp and hasp plate

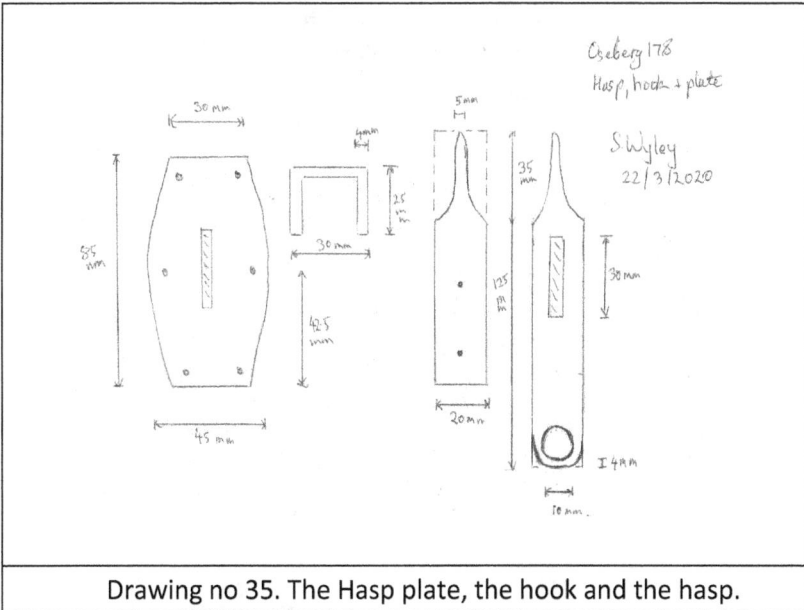

Drawing no 35. The Hasp plate, the hook and the hasp.

Construction Method for making hasp and hasp plate.

Hasp (like hinge but with one piece with one end has an eye the other is the hook.)

Use the loop of the hasp plate to mark out the hole for the loop.

Drill multiple holes along the length of hasp to match the loop.

Cold punch the remaining metal between the holes.

File out the remains to fit the loop.

Hasp plate.

1. Cut to size.

2. Drill nail holes

3. Cut and bend loop.

4. Drill loop holes.

5. Rivet [32]loop in place Some chests are secured with just hasp and hasp plate with the addition of a padlock, such as the Voxtorp chest.[33]

A hasp is similar to a hinge but the hasp has a hole so the loop of the hasp plate can pass through and the padlock hook is then passed through the loop and the padlock is closed. The size and placement of the hole in the hasp is dictated by the size and shape of the hasp plate loop and the placement of the hasp plate.

---

[32] Riveting involves the peen over of the end of a piece of metal to hold it in place.

[33] From the Voxtorp Church, Småland, Sweden, c. 1200, Statens Historiska Museum, Stockholm, 4094.

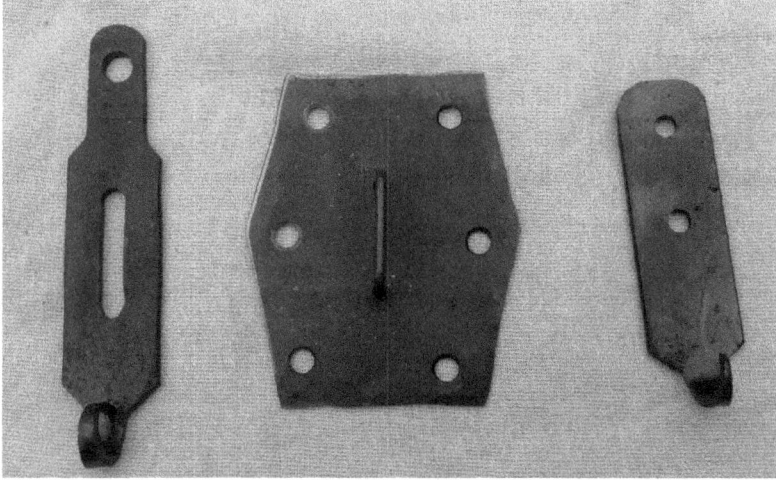

Photo 73. Hasp and hasp plate.

Photo 74. Hasp and hasp plate attached to a chest.

Attaching – Hook, Hinge and staple.

Turn chest on its back and open lid to expose nails, hinge and staple.

Place two pieces of wood under each end of the chest.

Place lump of steel as thick at the pieces of wood under each end of the chest, under the nails, hinge and staple.

Bend the nails, hinge and staple over with a pair of pliers in opposite directions.

Hit bent nails, hinge and staple so they form a staple into the wood of the hinge.

Attaching – Hasp Plate.

Turn chest on its front and open lid to expose the nails.

Place two pieces of wood under each end of chest.

Place lump of steel as thick at the pieces of wood under each end of the chest under the nails.

Bend the nails over with a pair of pliers in opposite directions.

Hit bent nails so they form a staple into the wood of the hinge.

Appendix 3 – A longer form of the Oseberg bed

based on timber available at the local timber merchant.

Materials.

Table 18. Material details.		
Pine	Measurement per item.	Total amount of material required.
2 Sides	210cm * 29cm * 2cm	420cm
2 ends	180cm * 20cm * 2cm	180 cm
4 legs	58 * 7cm * 7cm	212cm
Rope	17 metres of at least 8 mm rope	

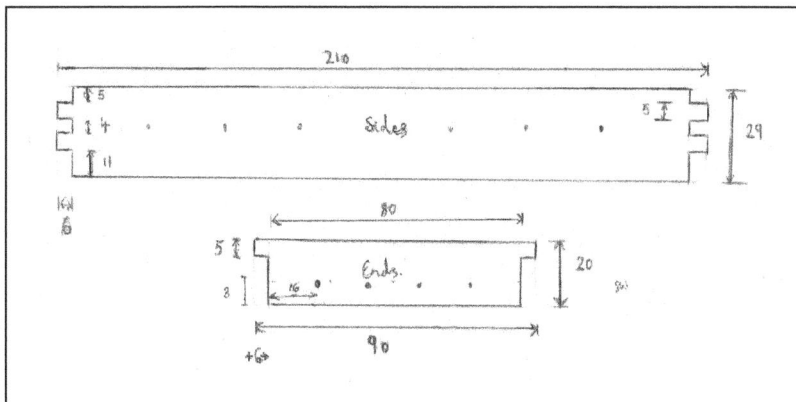

Drawing 36. The sides.

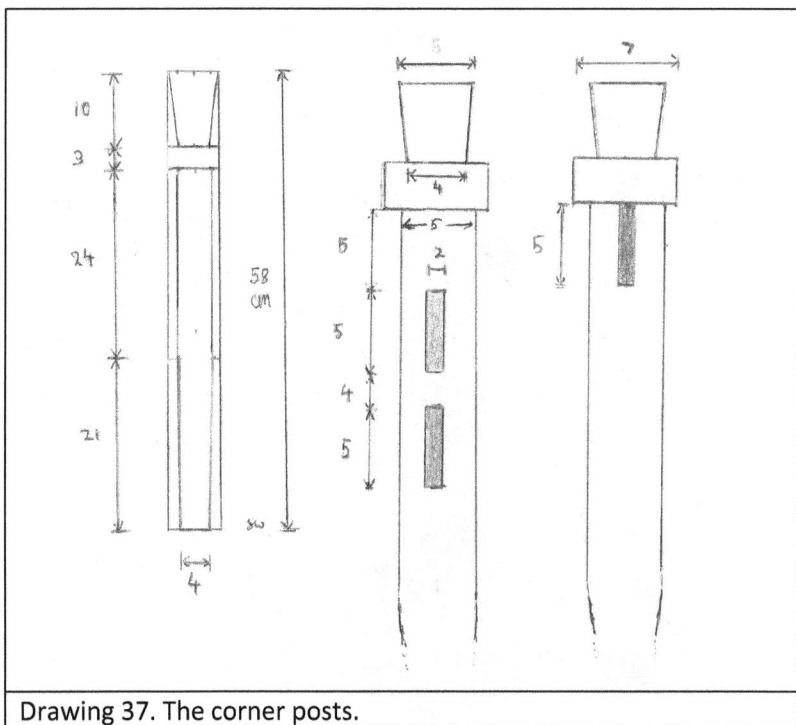

Drawing 37. The corner posts.

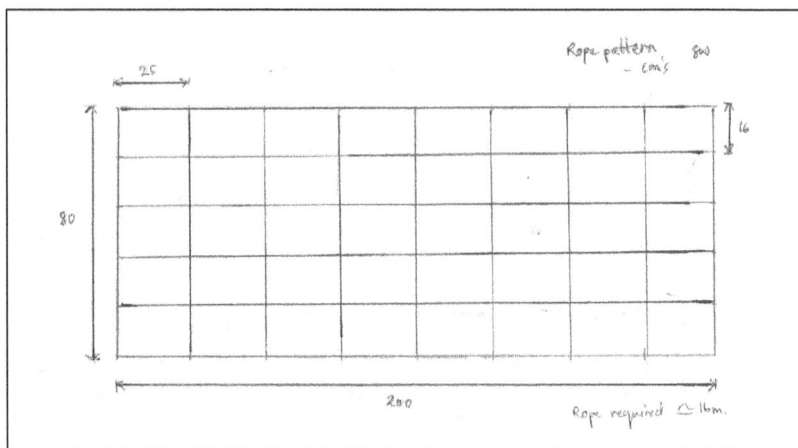

Rope pattern, 8w
- cm's

25

16

80

200

Rope required ≈ 16m.

Drawing 38. The suggested roping pattern, run the long side first, then weave in the short side into the long side.

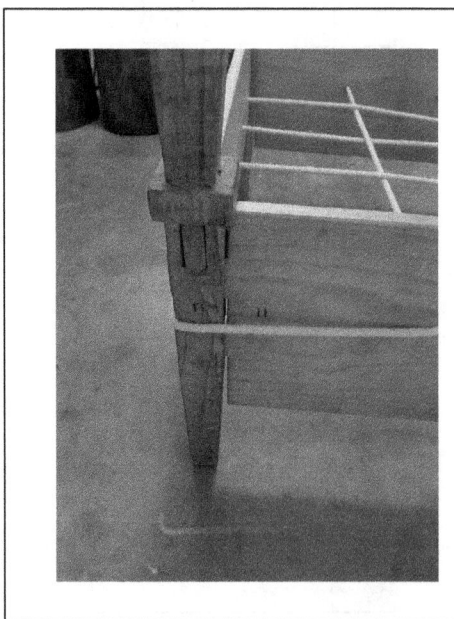

Photo 75. Showing the roping going around one of the legs, and a leg and side marked up with "Marriage Marks"[34] in Roman numerals to aid in assembly of the bed.

---

[34] https://moresuntimberframes.com/marriage-marks/ down loading 11/5/2022.

About The Authors

Stephen Francis (Sven) Wyley.

Born in 1962 in the town of Colac (Victoria, Australia) and now lives in Melbourne (Victoria), son of a nurse and a fibrous plaster, worked in analytical and research science (from EPA to CSIRO) for 25 years, then moved in occupational health and safety because safety paid more and he had desire to make the world a safer place one workplace at a time, and now works for the state government as a safety inspector.

Sports and martial arts figured predominantly in his life which included Freestyle wrestling (represented Australia), Archery (Longbow, Victorian State champion and competed at the nationals), Fencing (foil, epee and sabre at intervarsity competitions), and attaining a blue belt in Shudokan Aikido.

A living historian and re-enactor since joining the New Varangian Guard Inc. in 1984, then started Sven the Merchant (a business making and selling medieval goods mainly consisting of furniture, arrows and leatherwork) in 2004.

In his spare time he trains actors how to fight with medieval weapons, provides fight choreography and props for theatre, television and film as a director of Actor Fight Training Australia, and to aid with understanding there has been many up loads of short films to YouTube about using and maintaining tools, and making furniture.

35

https://sites.google.com/site/svenskildbiter/
https://www.youtube.com/chan-
nel/UCYaAcfT70zRoQv2RzNC829A

---

[35] Stephen Francis Wyley, not to be confused with his cousin or an actor in Ireland of the same name.

Wayne Robinson

Wayne Robinson and a shoe horn.

Wayne Robinson is the product of an upbringing by three genera-
tions of his family. This goes a long way to explaining his fascination
with tools and with fire. Working as a Telecommunications Technical
Officer for 12 years in exchange install, and then optical fibre and

satellite communications research and development at OTC. Six years in Legal and professional publishing followed, developing and producing digital reference and compliance works. He then had the good fortune to be in the right place at the right time as the telecommunication and IT industries converged and spent the next 15 years in a variety of IT and infrastructure roles. Accidentally retired, Wayne is working on the Renaissance Person thing, although the sonnets are proving to be an embuggerance.

A re-enactor of some vintage, having joined 1066 in 1982 and going on to be a member at different times (and occasionally, founder) of more re-enactment groups than can possibly be healthy. He is currently a member of The Pike and Musket Society, concentrating on the middle class London militia of 13 November, 1642.

He is sufficiently skilled in ancient, medieval and early modern period leather work, fletching, paint, woodwork and metalwork to bluff convincingly. Much to his own surprise, he finds himself to be a world authority on the carved hornwork of Robert Mindum (active 1593-1613).

Publications include: *The Reverend's Big Book of Leather* (2005) and *The Reverend's Big Blogge of Leather* (2009 -); *A Catalogue of Shoehorns by Robert Mindum* (2014 -); *The Opus of Robert Mindum- 1593 to 1613* (2018 -); with *The Double Armed Man - a New Invention* in preparation, based on a series of articles about William Neade's 1623 method of arming pikemen with longbows.

Link for the Reverend's blog.

https://leatherworkingreverend.wordpress.com/category/items/

Shannon Joyce

Shannon is a relative newcomer to the world of re-enacting. Her lifelong interest in history combined with her quest for a deeper understanding of her cultural roots has resulted in her diving head first into the Viking Age. Shannon has a passion for the material culture of this period with a focus on objects that would have been used daily in domestic life; especially those involved in the production and use of textiles and food preparation.

Shannon is an active member of Europa Re-enactment Association and when she's not planning her next craft project she works as a Data Analyst in the Public Health System. Shannon resides in the Hawkesbury/Blue Mountains north-west of Sydney with her adult children and cats.

https://www.facebook.com/KatlaKnarrabringa

Bibliography

Adams, R. (2005). Adhesive Bonding: Science, Technology and Applications, Elsevier.

Almgren, et al. (1991). The Viking, New York.

Almgren, O.(1907). Vikingatidssgrafvar I Sågan vid Sala, Forn Vännen 2-1-1907, Journal of Swedish Antiquarian Research.

Arwidsson, G. & Berg, G. (1983). The Mästermyr Find, A Viking Age Tool Chest from Gotland, Stockholm.

Bernstein, D. (1986). The Mystery of the Bayeux Tapestry, London.

Bill, Jan. (2011). Revisiting Gokstad Interdisciplinary investigations of a find complex excavated in the 19th century. https://www.researchgate.net/publication/281118137_Revisiting_Gokstad_Interdisciplinary_investigations_of_a_find_complex_excavated_in_the_19th_century, accessed 19 February 2020.

Blomquist, R., Martensson, A.W., (1963). Thulegrävningen 1961. En Berättelse om vad Grävninarna för Thulehuset i Lund Avslöjade, Archeologica Lundensia, volume 2.

Brisbane, Mark A. ED. (1992). The Archaeology of Novgorod, Russia, Recent results from the town and its hinterland, Lincoln.

Broderick, L. (2014). Satchels and bags in the early middle ages. Retrieved from Europa Reenactment Association Inc: http://europa.org.au/index.php/articles/21-bags, accessed 5 June 2021.

Croom, A.T. (2007). Roman Furniture, Great Britain.

Delort, R. (1973). Life in the Middle Ages, London.

Diehl, D. (1997). Constructing Medieval Furniture, Plans and Instructions with Historical Notes, Stackpole Books, USA.

Diehl, D & Donnelly, M. (1999). Medieval Furniture, Plans and Instructions for Historical Reproductions, Stackpole Books, USA.

Diehl, D & Donnelly, M. (2012). Medieval and Renaissance Furniture, Plans and Instructions for Historical Reproductions, Stackpole Books, USA.

Eklöf, N., (2017). Väskbyglarna från Birka / The purse frames from Birka. https://www.sjohistoriska.se/marinarkeologi3/marinarkeob-loggen/2017/vaskbyglarna-fran-birka--the-purse-frames-from-birka-english-translation-in-end-of-the-blogg, download 31/5/2020.

Gilbert, Vicenc & Lopez, Josep (2002). Woodworking class Cabinet-making, Pavilion Books.

Goodall, Ian H. (2012). Ironwork in Medieval Britain (Society for Medieval Archaeology Monographs, No. 31, UK.

Graham-Campbell, (1980). The Viking World, USA, 1980.

Graham-Campbell, James, (1989). The Viking World, Windward, London.

Graham-Campbell & Kidd, (1980). The Vikings, London, 1980.

Hawthorn & Smith, Trans. (1979). Theophilus, On Divers Arts, The Foremost Medieval Treatise on painting, glassmaking and metalwork, New York.

Heelas, Edgar H. (1944). Craftwork in Wood, Oxford University Press, Glasgow.

Horwood, Roger (2002). The Woodworker's Handbook, Caxton Publishing.

Iversen, M. (1991). Mammen. Grav, Kunst og samfund I vikingetid, Jysk Arckælogisk Selskabs Skrifter XXVII.

Jackson, Albert & Day, David, (1993). Collins Complete Wood Worker's Manual, London.

Jones, Peter (1987). Shelves, Closets & Cabinets, New York.

Kalmring, S. (2010). Of Thieves, Counterfeiters and Homicides. Crime in Hedeby and Birka. Fornvännen 105. Stockholm. http://www.fjellborg.org/Documents/Hedeby.pdf   downloaded 30/11/2020.

Killen, G. (1994). Egyptian Woodworking and Furniture, Shire Egyptology, UK.

Mercer, Henry, C. (reprint 2012). Ancient Carpenters' Tools: Illustrated and Explained, Together with the Implements of the Lumberman, Joiner and Cabinet-Maker in Use in the Eighteenth Century, New York.

Morris, Carole A. (2000). Archaeology of York: Craft, Industry and Everyday Life: Wood and Woodworking in Anglo-Scandinavian and Medieval York, v. 17, Fasc. 13.

Nicolaysen, (1882). Langskibet Fra Gokstad Ved Sandefjord, Cammermeyer, Christiana.

Ottaway, Patrick, (1992). Anglo-Scandinavian Ironwork from 16-22 Coppergate, Archaeology of York, Archaeology of York: Small finds, Part 6 of Small finds, York.

Raymond, C. (2017). Reflections on Birka and Hedeby Bags. Retrieved from Loose Threads: Yet Another Costuming Blog: http://cathyscostumeblog.blogspot.com/2017/05/reflections-on-birka-and-hedeby-bags.html, accessed 5 June 2021.

Roesdahl & Wilson, Eds, (1992). Viking to Crusader, The Scandinavians and Europe 800 -1200, Sweden.

Schleining, L. (2001). Treasure Chests, The Legacy of Extraordinary Boxes, The Taunton Press.

Stephan, Elizabeth A.; Park, William J.; Sill, Benjamin L.; Bowman, David R. & Ohland, Matthew W. (2010). Thinking Like an Engineer: An Active Learning Approach. Prentice Hall. p. 259. ISBN 0-13-606442-6.

UnitConverters.net, (2008 – 2020). Length Conversion, https://www.unitconverters.net/length-converter.html, accessed 5 March 2020.

Vlasaty, T. (2016). Bag-handles. Retrieved from Projekt Forlog: http://sagy.vikingove.cz/wp-content/uploads/2016/02/bag-handles , accessed 5 June 2021.

Watson, J. Et al. (2004). Townfoot Farm Cumwhitton, Cumbria. Investigation of material from the Viking Cementry.

Westphal, F. (2006). Die Holzfunde von Haithabu. Neumunster: Wachholtz Verlag.

Wilson,D., Ed., (1980). The Northern World, London.

Wyley, S., (2014) A Few Things Re-enactors can do to make there encampments better – Inspector General of the Encampment Inspectorate.

Further Reading

Abbott, M., (1991). Green Woodwork: Working with wood the Natural Way, Wiltshire.

Birkebaek, F., (1982). Danmarkshistorien, Vikingetiden, Kobenhavn.

Brisbane, M., Ed., (1992). The Archaeology of Novgorod, Russia, The Society of Medieval Archaeology, monograph 13.

Corkhill, T., (1980). The Complete Dictionary of Wood, Marboro Books.

Foley, D., (1962). Toys through the Age, Chilton.

Fridstrøm, E., (1985). The Viking Age Wood Carvers, Their Tools and Techniques, Univestetets Oldsaksamlings Skrifter, Oslo.

Johnson, H., (1980). The International Book of Wood, London.

Kolchin, B.A., (1989). Wooden Artifacts from Medieval Novgorod, Oxford.

Lang, J.T., (1988). Viking Age Decorated Wood, Royal Irish Academy.

Larsdatter, Karen, Medieval and Renaissance Culture, Chests and Trunks.

http://www.larsdatter.com/chests.htm Downloaded 9/07/2022.

Olsen, O., (1977). Fyrkat: En jysk vikingeborg. 1977.

Oughton, F., (1985). Viking Carving: How to do it, Mansard Press, UK.

Werner, A., (1999). London Bodies The Changing Shape of Londoners from Prehistoric Times to the Present Day, Museum of London.

Links

Moresun Timber Frames (2018)

https://moresuntimberframes.com/marriage-marks/ down loading 11/5/2022

Experimental Archeology (2022).

https://exarc.net/experimental-archaeology down loaded 29/06/2022

Photographs

Number	Description	Credit
Table 19. Photographs		
1	Trondheim toy horse replica by S.Wyley.	Morgan Wyley
2	Some forged nails by S.Wyley.	Stephen Wyley
3	Period tools, some of the tools pictured could be found in the Mästermyr tool box and other Viking tools.	Bryn Wyley
4	Hook knife marks on a replica of the Nydam quiver by W.Robinson.	Wayne Robinson
5	Blackening chest fittings	StephenWyley
6	Blackening chest fittings. The carbonizing of the oil.	StephenWyley
7	Chest made by S.Wyley based on the Cantigas de Santa Mari´a, Jewish merchants, Library of El Escorial. Madrid. Spain. National Heritage.	Joshua Button
8	Hook and eye hinges by Driffa Armories.	Stephen Wyley
9	Stephen Wyley with his Mästermyr chest.	Morgan Wyley
10	Barrel padlocks and keys by Matuls.	Stephen Wyley
11	Wooden mallet by S.Wyley.	Stephen Wyley
12	The seat of the Lund Viking stool. 11th century (Kulturen, Lund, Sweden, KM53. 436).	Kulturen, Lund, Sweden.

13	Lund Stool replica by S.Wyley.	Stephen Wyley
14	Drill, hole saw and jig used to cut angled leg holes.	Stephen Wyley
15	Drill, hole saw and jig used to cut first angled leg hole.	Stephen Wyley
16	Oseberg 178 chest.	Viking ship museum, Oslo, Norway.
17	Oseberg 178 chest with lock by G.Baker and S.Wyley.	Bryn Wyley
18	Oseberg 178 chest bottom tenon marked out.	Stephen Wyley
19	Oseberg 178 chest bottom, tenons cut out.	Stephen Wyley
20	Oseberg 178 chest end, sides cut off.	Stephen Wyley
21	Oseberg 178 chest end, sids of internal halving joint cut.	Stephen Wyley
22	Oseberg 178 chest end, internal halving joint chiselled out.	Stephen Wyley
23	Oseberg 178 chest end, mortice chiseled out.	Stephen Wyley
24	Oseberg 178 chest bottom presenting tenon for ends to be fitted.	Stephen Wyley
25	Oseberg 178 chest bottom end fitted on bottom.	Stephen Wyley
26	Oseberg 178 chest, marking and ensuring the angle of the front and back are the same.	Stephen Wyley
27	Oseberg 178 chest, marking and ensuring the angle of the front and back are the same.	Stephen Wyley

28	Oseberg 178 chest, placement of nails compared to the halving joint.	Stephen Wyley
29	Oseberg 178 chest, Placement and skewing of the nails.	Stephen Wyley
30	Oseberg 178 chest, the box of the chest.	Stephen Wyley
31	Oseberg 178 chest, the lid on the chest.	Stephen Wyley
32	Oseberg 178 chest, the hinges and staples.	Stephen Wyley
33	Oseberg 178 chest, Adding the top of the hinge.	Stephen Wyley
34	Oseberg 178 chest, hinge loop in the back of the lid and held in place by the Loop staple.	Stephen Wyley
35	Oseberg 178 chest, Hinge with all 3 staples.	Stephen Wyley
36	Oseberg 178 chest, the cleating on the inside of the chest for a hinge.	Stephen Wyley
37	Oseberg 178 chest, the spring, the lock plate and the key.	Stephen Wyley
38	Oseberg 178 chest, the lock in place on the chest.	Stephen Wyley
39	Oseberg bed no. 192	Viking ship museum, Oslo, Norway. http://www.uni-mus.no/foto/imageviewer.html#/ ?id=1735063&type=jpeg

40	Oseberg bed , Replica by S.Wyley	Stephen Wyley
41	Oseberg bed no.2, Horizontal saw cuts on the leg.	Stephen Wyley
42	Oseberg bed no.2, Diagonal cuts down to the bead on the leg.	Stephen Wyley
43	Oseberg bed no.2, Re-enactor Ben Johnson from the New Varangian Guard Inc. – Vlachernai branch shaving legs down to size.	James Brown
44	Oseberg bed no.2, Legs shaved down to size, tenons next.	Stephen Wyley
45	Oseberg bed no.2, Leg on the bench secured by a 'hold down' ready to start chiseling.	Stephen Wyley
46	Oseberg bed no.2, Tenons in the legs are done.	Stephen Wyley
47	Oseberg bed no.2, Legs fitted to sides.	Stephen Wyley
48	Oseberg bed no.2, Legs fitted to ends.	Stephen Wyley
49	Oseberg bed no.2, The longer version of the bed in pine, with rope support for mattress.	Kym Tobin
50	Sala Hytta table, top of table, showing the chiselled out section the tops of the for legs.	Stephen Wyley
51	Sala Hytta table, the table from the corner of side of table.	Stephen Wyley
52	Sala Hytta table, drilling holes for legs with drill, spade bit and jig.	Stephen Wyley
53	Sala Hytta table, chiselling out the insert in the table top.	Stephen Wyley

54	Sala Hytta table, marking the leg.	Stephen Wyley
55	Sala Hytta table, cutting the tenon.	Stephen Wyley
56	Sala Hytta table, chiselling the tenon.	Stephen Wyley
57	Sala Hytta table, rounding off the tenon with a rasp.	Stephen Wyley
58	Sala Hytta table, fitting the legs.	Stephen Wyley
59	Trondheim toy horse.	Vitenskapsmuseet Trondheim, Norway.
60	Trondheim toy horse replica by S.Wyley.	Stephen Wyley
61	Trondheim toy horse replica, the tail.	Stephen Wyley
62	Trondheim toy horse replica, the ears and eyes.	Stephen Wyley
63	Trondheim toy horse replica, the mane.	Stephen Wyley
64	Gokstad Candle stand.	Viking ship museum, Oslo, Norway.
65	Gokstad Candle stand by W.Robinson	Wayne Robinson
66	Gokstad Candle stand by W.Robinson	Wayne Robinson
67	Birka Bag topper	Swedish National Maritime and Transport

		Museums, Stockholm, Sweden.
68	Completed hanger based on Birka F153	Shannon Joyce
69	Sami bag by Norske Folkmuseum	Free use via a creative commons licence deed by SA 4.0.
70	Template of Birka bag hanger F153	Shannon Joyce
71	Completed hanger.	Shannon Joyce
72	Completed bag.	Shannon Joyce
73	Hasp and hasp plate for an Oseberg 178 chest.	Stephen Wyley
74	Hasp and hasp plate attached to a chest.	Stephen Wyley
75	Gokstad bed corner post and roping.	Kym Tobin
76	Stephen 'Sven' Wyley	Robbie Last
77	Wayne Robinson	Glenda Robinson
78	Shannon Joyce	J.P.Harris

Table 20. Drawings.		
Number	**Description**	**Credit**
1	Joints of the Mästermyr chest	Steven Lowe
2	Coppergate mallet no. 8186	Morris, Carole A. (2000). *Archaeology of York: Craft, Industry and Everyday Life: Wood and*

		Woodworking in Anglo-Scandinavian and Medieval York, v. 17, Fasc. 13.
3	Coppergate mallet no. 8186	Dimensions added Morris drawing.
4	Lund stool – top view	Jenny Baker.
5	Lund stool – front view	Jenny Baker
6	Lund stool – side view	Stephen Wyley
7	Oseberg chest 178 - plans	Stephen Wyley
8	Oseberg chest 178 – top part of hinge	Stephen Wyley
9	Oseberg chest 178 – Loop staple	Stephen Wyley
10	Oseberg chest 178 - staple	Stephen Wyley
11	Oseberg chest 178 – hinge and how it connects the lid to the back of the chest, side view.	Jenny Baker
12	Oseberg chest 178 – Lock (plate, spring and key)	Jenny Baker
13	Oseberg chest 178 – Lock cut out information	Jenny Baker
14	Oseberg bed no.192	Unknown
15	Oseberg bed no.192 – side and end rails	Stephen Wyley
16	Oseberg bed no.192 – Corner post	Stephen Wyley

17	Oseberg bed no.192 – roping pattern for mattress support	Stephen Wyley
18	Sala Hytta table - showing top and side views	Unknown
19	Sala Hytta table - overhead view	Stephen Wyley
20	Sala Hytta - side views	Stephen Wyley
21	Trondheim Toy Horse	Photo with dimensions by Stephen Wyley
22 - 27	Gokstad Candle holder	Wayne Robinson
28	Hedeby of bag hangers with textile fragments. Source: Schietzel, (2014) as cited in Vlasaty, (2016)	Schietzel
28 - 33	Bag topper	Shannon Joyce
34	Plans for a larger version of the Oseberg 178 chest	Stephen Wyley
35	Plans for a Hasp and hasp plate set for an Oseberg 178 chest	Stephen Wyley
36	Oseberg bed no.192 – the sides.	Stephen Wyley
37	Oseberg bed no.192 – the corner posts.	Stephen Wyley
38	Oseberg bed no.192 – the roping pattern.	Stephen Wyley

	Table 21. Illuminated Manuscripts.
Number	Description
1	Cantigas de Santa Mari'a (Canticles of Holy Mary). Reign of Alfonso X of Castile, "the Wise" (1221-1284). Jewish merchants, Library of El Escorial. Madrid. Spain, National Heritage.
2	Tacuninum sanitatis, 47v., detail. Late 14th century. Bibliotheque Nationale de France.
3	Tacuninum sanitatis, 29v., detail. Late 14th century. Bibliotheque Nationale de France.
4	The disrobing of Christ from the Karlsruhe Passion by Hans Hirtz c.1440. One of the 6 panels held in the Staatliche Kunsthalle (State Art Gallery) in Karlsruhe, Germany.
5	Detail of one of the frescos to be found in the crypt of the Saint-Nicolas Church in Tavant, France. Dating from the 12th century. Before Chartres: Appunti sull'arte romanica e sul tempo romanico https://beforechartres.blog/2018/04/06/i-capo-lavori-nelle-viscere-di-tavant/
6	Harley Psalter 66v. First half of the 11th century. British Library http://www.bl.uk/manuscripts/Viewer.aspx?ref=harley_ms_603_fs001r

Index

Forthcoming Volumes.

14th century.
- The Chair - Bede' Chair (St.Paul's Church, Jarrow, County Durham, United Kingdom);
- The Chest and lock - The Pillager's chest (MS Royal manuscript, British Library, UK);
- The Bed - The Sitting bed (The Preparation of the Cross , Sant Croce Fresco, Florence, Italy);
- The Table - Making Pasta (Taciunum Sanitatis of Vienna manuscript, BNF, France);
- The Table (The Luttrel Psalter Table);
- The Toy Sword ;
- The Kitchen item - The Hanging Salt Box (Budapest Historical Museum, Hungary);
- The Tool - The Bellows (Smithfield Decretals manuscript, British Library, UK);
- The Tool - The Frame Saw (The Sant Croce Fresco, Florence, Italy);
- The Clothing accessory - the Aglet;

The Leather work - The Costrel (Museum of London, UK);
The Heater Shield.

Next Volumes:
Viking Volume 2;
Byzantine;
Crusader (11 – 13th century);
15th century & 16th century.